# Medical gases
# Health Technical Memorandum 02-01: Medical gas pipeline systems

## Part A: Design, installation, validation and verification



London: The Stationery Office

Published by TSO (The Stationery Office) and available from:

**Online**
**www.tsoshop.co.uk**

**Mail, Telephone, Fax & E-mail**
TSO
PO Box 29, Norwich NR3 1GN
Telephone orders/General enquiries 0870 600 5522
Fax orders 0870 600 5533
E-mail customer.services@tso.co.uk
Textphone 0870 240 3701

**TSO Shops**
123 Kingsway, London WC2B 6PQ
020 7242 6393 Fax 020 7242 6394
68–69 Bull Street, Birmingham B4 6AD
0121 236 9696 Fax 0121 236 9699
9–21 Princess Street, Manchester M60 8AS
0161 834 7201 Fax 0161 833 0634
16 Arthur Street, Belfast BT1 4GD
028 9023 8451 Fax 028 9023 5401
18–19 High Street, Cardiff CF10 1PT
029 2039 5548 Fax 029 2038 4347
71 Lothian Road, Edinburgh EH3 9AZ
0870 606 5566 Fax 0870 606 5588

**TSO Accredited Agents**
(see Yellow Pages)
*and through good booksellers*





ISBN 0-11-322742-6
978-0-11-322742-6

First published 2006

Printed in the United Kingdom for The Stationery Office

The paper used in the printing of this document (Revive Silk) is 75% made from 100% de-inked post-consumer waste, the remaining 25% being mill broke and virgin fibres. Recycled papers used in its production are a combination of Totally Chlorine Free (TCF) and Elemental Chlorine Free (ECF). It is recyclable and biodegradable and is an NAPM and Eugropa approved recycled grade.

# Preface

## About Health Technical Memoranda

Engineering Health Technical Memoranda (HTMs) give comprehensive advice and guidance on the design, installation and operation of specialised building and engineering technology used in the delivery of healthcare.

The focus of HTM guidance remains on healthcare-specific elements of standards, policies and up-to-date established best practice. They are applicable to new and existing sites, and are for use at various stages during the whole building lifecycle:

*Figure 1 Healthcare building life-cycle*

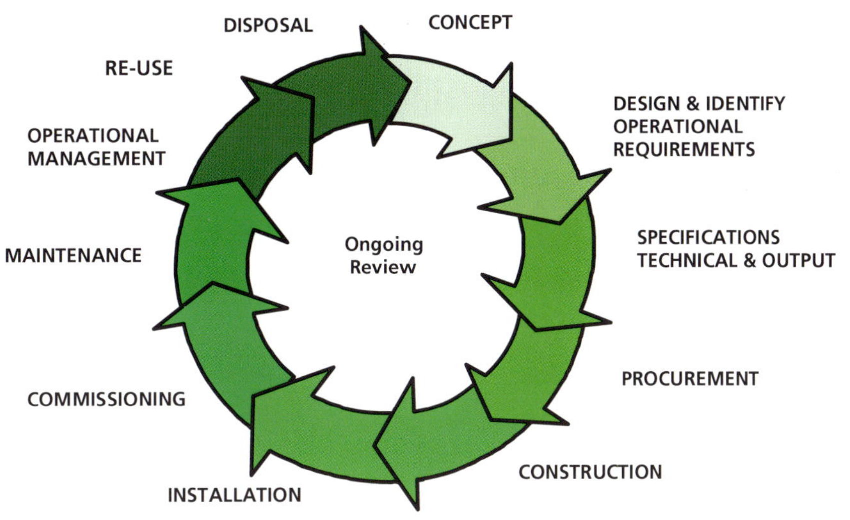


Healthcare providers have a duty of care to ensure that appropriate engineering governance arrangements are in place and are managed effectively. The Engineering Health Technical Memorandum series provides best practice engineering standards and policy to enable management of this duty of care.

It is not the intention within this suite of documents to unnecessarily repeat international or European standards, industry standards or UK Government legislation. Where appropriate, these will be referenced.

Healthcare-specific technical engineering guidance is a vital tool in the safe and efficient operation of healthcare facilities. Health Technical Memorandum guidance is the main source of specific healthcare-related guidance for estates and facilities professionals.

The new core suite of nine subject areas provides access to guidance which:

- is more streamlined and accessible;
- encapsulates the latest standards and best practice in healthcare engineering;
- provides a structured reference for healthcare engineering.

## Structure of the Health Technical Memorandum suite

The new series of engineering-specific guidance contains a suite of nine core subjects:

Health Technical Memorandum 00
Policies and principles (applicable to all Health Technical Memoranda in this series)

Health Technical Memorandum 01
Disinfection and sterilization

Health Technical Memorandum 02
Medical gases

Health Technical Memorandum 03
Ventilation systems

Health Technical Memorandum 04
Water systems

Health Technical Memorandum 05
Fire safety

Health Technical Memorandum 06
Electrical services

Health Technical Memorandum 07
Environment and sustainabilty

Health Technical Memorandum 08
Specialist services

Some subject areas may be further developed into topics shown as -01, -02 etc and further referenced into Parts A, B etc.

Example: Health Technical Memorandum 06-02 Part A will represent:

Electrical Services – Safety – Low Voltage

In a similar way Health Technical Memorandum 07-02 will simply represent:

Environment and Sustainability – $EnCO_2de$.

All Health Technical Memoranda are supported by the initial document Health Technical Memorandum 00 which embraces the management and operational policies from previous documents and explores risk management issues.

Some variation in style and structure is reflected by the topic and approach of the different review working groups.

DH Estates and Facilities Division wishes to acknowledge the contribution made by professional bodies, engineering consultants, healthcare specialists and NHS staff who have contributed to the review.

*Figure 2 Engineering guidance*

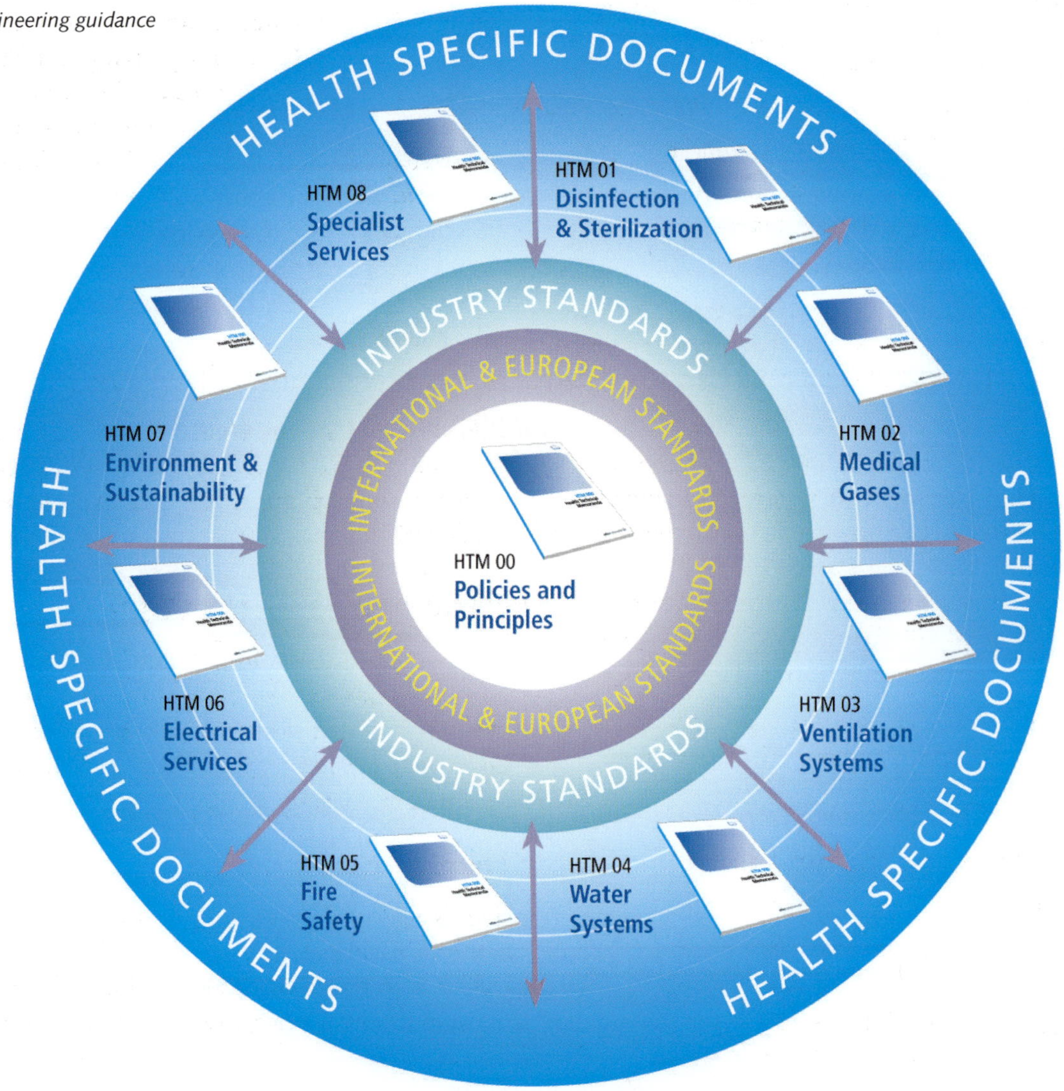

# Executive summary

## Introduction

A medical gas pipeline system (MGPS) is installed to provide a safe, convenient and cost-effective system for the provision of medical gases to the clinical and nursing staff at the point-of-use. It reduces the problems associated with the use of gas cylinders such as safety, porterage, storage and noise.

This Health Technical Memorandum is divided into two parts. Guidance in this part (Part A) covers piped medical gases, medical and surgical air, and medical vacuum installations: it applies to all medical gas pipeline systems installed in healthcare premises and anaesthetic gas scavenging disposal systems. Specifically, it deals with the issues involved in the design, installation, and validation and verification (testing and commissioning) of an MGPS. Part B covers operational management.

The guidance given in this document should be followed for all new installations and refurbishment or upgrading of existing installations.

It is not necessary to apply the guidance retrospectively unless patient or staff safety would be compromised. In this case, the guidance given in this document should be followed.

Existing installations should be assessed for compliance with this guidance document. A plan for upgrading the existing system should be prepared, taking account of the priority for patient safety. Managers will need to liaise with medical colleagues and take account of other guidance published by the Department of Health in order to assess the system for technical shortcomings.

Health Technical Memorandum 02 supersedes all previous versions of Health Technical Memorandum 2022.

## Sources of supply for pipeline installations

### Oxygen

Oxygen is generally supplied from:

- a liquid source such as a large vacuum-insulated evaporator (VIE);
- liquid cylinders or compressed gas cylinders; or
- a combination of these to provide the necessary stand-by/back-up capacity.

Oxygen can also be supplied from an oxygen concentrator (pressure-swing adsorber). Such systems are usually installed where liquid or cylinders are expensive, unavailable or impracticable.

### Medical air

Medical air is usually supplied from a compressed air plant that includes high-quality drying and filtration equipment. Blending oxygen and nitrogen on-site to provide a high-quality product with minimum maintenance can also provide medical air. Where such systems are installed to provide both oxygen and medical air, nitrogen can be used for the power source for surgical tools.

### Other gases

All other gases are supplied from cylinders.

(On-site blended oxygen/nitrous oxide mixture is a possibility if bulk liquid supplies of nitrous oxide are available, although this system is unlikely to be adopted in the UK.)

## Basic principles of design

Patient safety is paramount in the design, installation, commissioning and operation of medical gas pipeline systems. The basic principles of safety are achieved by ensuring quantity of supply, identity of supply, continuity of supply and quality of supply.

### Quantity of supply

This is achieved by ensuring that the design of the pipeline installation and capacity of the supply plant is sufficient to provide the required flows of gases and vacuum for the intended number of patients to be treated at any one time. Adequacy of supply is established during commissioning of the systems.

### Identity of supply

This is achieved by ensuring that all points to which the user can connect medical equipment (terminal units) and user-replaceable components are provided with gas-specific connectors. Such connectors are also identified by symbol and often colour. The gas specificity is maintained by comprehensive tests and checks during installation and commissioning, and during any work or maintenance on the systems.

### Continuity of supply

This is achieved by installing, as a minimum, duplex components and providing additional means of supply provision in the event of failure of the primary and secondary plant or supply system. Systems are also connected to the essential electrical supply.

### Quality of supply

Quality of supply is ensured by the use of gaseous or liquid sources that are provided to an appropriate product specification, usually a recognised European Pharmacopoeia (Ph. Eur.) monogram. In the case of compressor-based systems, filtration equipment to a known and agreed standard is installed. To ensure that the product is not adulterated in the distribution system, pipeline installations and components are required to meet agreed specifications. There are strict Ph. Eur. requirements for medical gases.

## General uses of gas and pipeline installations

- Oxygen is one of the most extensively used gases for respiratory therapy and life-support and is additionally used in anaesthetic procedures.
- Medical air is mainly used in respiratory therapy as a power source for patient ventilators, and for blending with oxygen. It is also used as the driving gas for nebulised drugs and chemotherapy agents.
- Surgical air (of medical air quality) is also used, at a higher pressure, to power a variety of surgical tools and other devices such as tourniquets. (As an alternative, nitrogen can be used for this purpose.).
- Nitrous oxide is used for anaesthetic and analgesic purposes, being mixed with air, oxygen, and nebulised agents.
- Pipeline systems for a 50% mixture of oxygen and nitrous oxide are widely installed in the UK for analgesic purposes, particularly in maternity departments.
- Helium/oxygen mixture is used to treat patients with respiratory or airway obstruction and to relieve symptoms and signs of respiratory distress; guidance on pipeline systems is now included.
- Carbon dioxide is used less commonly now as a respiratory stimulant, and for insufflation during surgery. Pipeline systems for respiratory use have not been installed in the UK but they are now being installed for this latter purpose.
- Piped vacuum is provided in most clinical areas by means of centrally sited vacuum pumps.
- The control of occupational exposure to waste anaesthetic gas (nitrous oxide) and nebulised agents is a legal requirement under the Control of Substances Hazardous to Health (COSHH) Regulations 2002. Where nitrous oxide is provided for anaesthetic purposes, scavenging systems are installed.

# Acknowledgements

**Mike Arrowsmith** Arrowsmith and Associates

**Geoffrey Dillow** Geoffrey Dillow and Associates

**Ian Fraser** Department of Health

**Mike Ralph** Medical Gas Association

# Contents

Preface
Executive summary
Acknowledgements

**Chapter 1** Scope page 1
Guidance in this document
Other guidance

**Chapter 2** General principles page 2
Introduction
Quality requirements for medical gases and air
Sources of supply
Sizing information for gas supply sources
Pipeline distribution system design
Safety
Installation/supply of equipment/maintenance
Modifications
Removal of pipework
Validation and verification
General fire precautions
Electricity supply to medical gas installations

**Chapter 3** Provision of terminal units, and the location of AVSUs, local alarm indicator panels and LVAs page 10
General
Terminal units
Terminal units for helium/oxygen mixture
Nitrogen for surgical tools
AVSUs
Local alarm indicator panels
LVAs
Specific labelling requirements

**Chapter 4** Gas flow page 22
General
Terminal unit flows
Pipeline flows
Oxygen
Hyperbaric oxygen chambers
Nitrous oxide
Nitrous oxide/oxygen mixture
Air
Vacuum
Helium/oxygen mixture
Anaesthetic gas scavenging systems

**Chapter 5** Cylinder manifold installations page 36
- Primary supply system
- Secondary supply system

**Chapter 6** Oxygen systems page 41
- Liquid oxygen systems
- Oxygen concentrator installations (PSA plant)

**Chapter 7** Medical compressed air systems page 66
- General
- Compressor systems
- Quality
- Siting
- Compressor noise
- Air intake
- Compressor types
- Compressor lubrication
- After-coolers
- Receivers
- Air treatment and filtration
- Pressure control
- Traps, valves and non-return valves
- Operating and indicating system
- Synthetic air
- System description
- Storage vessels
- Vaporisation
- Medical oxygen flow control
- Surgical nitrogen flow control
- Control panel for the nitrogen and oxygen supplies to the mixing panels
- Air mixing panels
- Buffer vessels
- Alarm signal status unit
- Emergency supply provision
- Additional use of medical air systems

**Chapter 8** Surgical air systems page 82
- General
- Extension of surgical air systems into dental departments

**Chapter 9** Medical vacuum systems page 85
- General
- Siting
- Pump noise
- Vacuum plant exhaust
- Efficiency
- Vacuum pumps
- Pressure control
- Valves
- Pressure regulation of vacuum system
- Vacuum indicators
- Electrical supply
- Pump operating and indicating system

**Chapter 10** Anaesthetic gas scavenging disposal systems page 91
- Terminology
- General
- Selecting the number of disposal system pumps
- Flow and diversity
- Discharge outlet
- Plant control indication

**Chapter 11** Other medical gas pipeline installations page 94
- General
- Helium/oxygen mixture
- Oxygen/$CO_2$ mixture
- Carbon dioxide
- Nitric oxide

**Chapter 12** Warning and alarm systems page 96
- General
- Dedicated systems
- Panel location
- System components
- System layout
- General requirements
- Warning and alarm system faults
- Indicator panel requirements for all systems
- Central indicator panel requirements
- Repeater indicator panel requirements
- Area warning and alarm panel
- Integrated systems

**Chapter 13** Pipeline installation page 104
- Pipeline materials
- Cleaning
- Pipeline jointing
- Inspection of joints
- Jointing methods (mechanical)
- Capping
- Pipe supports
- Identification of pipelines
- Pipeline components
- Medical supply units
- Flexible pendant fittings
- Bed-head trunking/walling systems
- LVAs and AVSUs
- Specific labelling requirements
- Pressure control equipment
- Pressure sensors
- Pressure gauges
- Test points
- Emergency inlet port
- Line pressure alarms and safety valves
- Other systems

**Chapter 14** Accommodation page 114
- Design and construction of plant and manifold rooms

**Chapter 15** Validation and verification page 117
Summary of tests
General requirements for testing
Modifications, extensions or repairs to existing systems
Requirements for pipeline carcass tests
Labelling and marking
Sleeving and supports
Leakage
Cross-connection
Requirements for pipeline system tests
Leakage from total compressed medical gas systems
Leakage into total vacuum systems
Closure of area valve service units and line valve assemblies
Zoning of AVSUs and terminal unit identification
Cross-connection
Flow and pressure drop at individual terminal units, mechanical function and correct installation
Performance tests on the pipeline system
Functional tests of supply systems
Pressure safety valves
Warning and alarm systems
Verification of as-fitted drawings
Filling with medical air
Purging and filling with specific gases
Pharmaceutical testing
Quality of medical gas systems
AGS disposal systems
Requirements before a medical gas pipeline system is taken into use

**Appendix A** Testing, commissioning and filling for use: forms to be completed during testing and commissioning of piped medical gases systems page 141

**Appendix B** Gas pressure variation with temperature page 161
General
Calculation
Examples

**Appendix C** Pressure-drop test device page 162
General
Measurement principle
Functional requirements
Orifices
Flowmeter
Pressure gauge

**Appendix D** Membrane filter test device page 163
General
Measurement principle
Test equipment

**Appendix E** Equipment for contaminant testing page 164
General
Measurement principle

| | | |
|---|---|---|
| **Appendix F** | Equipment for gas identification<br>General<br>Specificity<br>Specification | page 165 |
| **Appendix G** | Pressure loss data<br>Pipeline pressure-drop calculations | page 166 |
| **Appendix H** | Checklist for planning/installing/upgrading a cryogenic liquid supply system<br>Delivery frequency<br>Calculating consumption<br>Age of current system<br>Siting of system and the site survey<br>Costs<br>Emergency provision<br>System shutdown during installation<br>Paperwork<br>Health and safety<br>Preparation<br>Installation<br>Follow-up | page 172 |
| **Appendix J** | Upgrading surgical air systems<br>Background<br>Modifying "old" systems | page 176 |
| **Appendix K** | Signage requirements | page 178 |
| **Appendix L** | Important notes for use of medical vacuum and anaesthetic gas scavenging<br>Infectious disease units<br>Laser/surgical diathermy smoke extraction<br>Dental vacuum systems<br>Anaesthetic gas scavenging (AGS) | page 182 |
| **Appendix M** | Oxygen usage data | page 183 |
| **Appendix N** | Pressure conversion table | page 184 |
| **References** | | page 185 |

# 1 Scope

## Guidance in this document

1.1 This Health Technical Memorandum is divided into two parts. Guidance in this part (Part A) covers pi[illegible]edical gases, medical and surgical air, and medical vacuum installations; it applies to all medical gas pipeline systems installed in healthcare premises. Anaesthetic gas scavenging disposal systems are also included. Specifically, it deals with the issues involved in the design, installation, and validation and verification (testing and commissioning) of an MGPS. Part B covers operational management.

1.2 The guidance given in this document should be followed for all new installations and refurbishment or upgrading of existing installations.

1.3 It is not necessary to apply the guidance retrospectively unless patient or staff safety would be compromised. In this case, the guidance given in this document should be followed.

1.4 Existing installations should be assessed for compliance with this guidance document. A plan for upgrading the existing system should be prepared, taking account of the priority for patient safety. Managers will need to liaise with medical colleagues and take account of other guidance published by the Department of Health in order to assess the system for technical shortcomings.

1.5 Throughout this document, "medical gas pipeline system(s)" will be described by the term MGPS.

## Other guidance

1.6 Model Engineering Specification C11 – 'Medical gases' supports this Health Technical Memorandum. It provides details of the extent of the works required and is a procurement specification.

1.7 Whenever appropriate, British Standard specifications should be used.

1.8 Guidance on the provision of MGPS is given in the Health Building Notes and other relevant British, European, and International standards.

# 2 General principles

## Introduction

2.1 An MGPS is designed to provide a safe and effective method of delivering medical gases, medical air and surgical air from the source of supply to the appropriate terminal unit by means of a pipeline distribution system. Medical vacuum is also provided by means of a pipeline system. Anaesthetic gas scavenging disposal systems are provided to control occupational exposure to waste anaesthetic gases and agents.

2.2 It is essential to ensure that there is no possibility of a cross-connection between any system and that all parts of each system to which connections can be made by users are gas-specific.

2.3 Dental compressed air and vacuum systems have differing requirements, and these are covered in Health Technical Memorandum 2022 Supplement 1 – 'Dental compressed air and vacuum systems'.

2.4 During the installation stage, extensive tests are carried out to verify that there is no cross-connection.

2.5 Medical gas systems may be extended to those departments where respiratory equipment or surgical tools are serviced, such as in electronic and biomedical equipment (EBME) workshops and sterile services departments (SSDs). Specific additional uses of air systems are covered in Chapters 7 and 8.

2.6 MGPS should not be used to supply pathology departments, general workshops or mechanical services.

2.7 Separate installations should be provided for pathology and general laboratories and workshops, although it is recommended that they be constructed to the same specification as MGPS. They should not be provided with medical gas terminal units.

### Note

Portable suction devices should be used in infectious disease units.

## Quality requirements for medical gases and air

2.8 Medical gases supplied from cylinder or liquid sources comply with the appropriate sections of the current edition of the European Pharmacopoeia (Ph. Eur.). The Ph. Eur. also specifies the approved testing methods to be adopted for gas identity.

2.9 The quality specification for medical, surgical and synthetic air, and oxygen-enriched air produced from a pressure swing adsorber (PSA) system, is as given in Table 29. The medical air and synthetic air should also comply with the appropriate sections of the current edition of the current edition of the Ph. Eur. (see Table 30).

2.10 The quality of piped medical compressed air, and the particulate content, dryness and concentration of impurities should comply with the requirements for maximum concentrations given in Table 30. Information on testing procedures is given in Chapter 15 "Validation and verification".

2.11 Bacteria filters should be included in medical and surgical compressor systems to reduce the risk of delivering spores or other infectious material to vulnerable patients.

2.12 Micro-organisms can penetrate a bacteria filter if the material is wet. Therefore it is essential that the dryness of the medical air supplied to a bacteria filter is checked regularly (at least every three months) at the test point, using the test equipment specified in Chapter 15.

## Sources of supply

2.13 Both BS EN 737-3:2000 and ISO 7396-1:2002 propose that all medical gas supplies should comprise three sources of supply identified as "primary", "secondary" and "reserve", although the latter is more commonly referred to as a third means of supply. The supply system should be designed to achieve continuity of supply to the terminal units in normal condition and in a single fault condition. A single fault condition is where a single means for protection against a safety hazard in equipment is defective or a single external abnormal condition is present. Loss of supply due to maintenance of a supply source (or a component within it) is not considered a single fault condition.

2.14 Comparing this Health Technical Memorandum with the above Standards will reveal a different classification of, for example, individual banks of a cylinder manifold. Whereas EN 737-3:2000 refers to the separate banks of an automatic manifold as primary and secondary supplies, this Health Technical Memorandum classifies such a manifold as a primary supply, that is, one single operating unit.

2.15 Regardless of these classification differences, the choice of central source will be defined by the ability of the source not only to provide a continuous supply of gas over a range of possible flow rates but also to offer security of supply by virtue of adequate capacity.

2.16 For these reasons, types, capacities and locations of primary, secondary and reserve sources of supply will be based on both system design parameters and the need for supply security, identified by a risk assessment during the planning stage. Security of medical air supplies must be given a high priority. Total electrical failure must not be allowed to jeopardise supplies, and all medical air systems must be supported by an appropriate fully-automatic manifold. Tables 1–9 describe the various options for gas supply. For each, the primary, secondary and reserve sources are identified.

### Table 1 Compressed gas cylinder manifold systems

| Primary supply | Secondary supply | Reserve supply (third source of supply) |
|---|---|---|
| Fully automatic manifold.<br>Number of cylinders based on system design | Manual emergency reserve manifold.<br>To come on line automatically via a non-return valve.<br>Number of cylinders based on ability to provide 4 hours' supply at average use | Automatic/manual manifold supplying via non-interchangeable screw thread (NIST) connectors<br>OR<br>Locally-based integral valved cylinders with regulators/flowmeters attached |

### Table 2 VIE systems

| Primary supply | Secondary supply | Reserve supply (third source of supply) |
|---|---|---|
| Simplex VIE (vacuum-insulated evaporator) vessel system | Automatic manifold system.<br>To come on-line in the event of plant failure | Automatic manifold system.<br>May be sited to support high-dependency areas or whole site<br>OR<br>Locally-based integral valved cylinders with regulators/flowmeters attached |
| One vessel of a duplex VIE (vacuum-insulated evaporator) vessel system (on same plinth) | Second vessel of a duplex VIE system | Automatic manifold system.<br>May be sited to support high-dependency areas or whole site |
| One vessel of a duplex VIE vessel system (on separate plinths) | Second vessel of a duplex VIE system (on separate plinths).<br>NB split-site systems are intended primarily for systems where the risk assessment has identified that the site for the primary supply is limited in size or presents too high a risk having both tanks on the same site. These supply systems should be fitted with appropriate non-return valved connections to prevent gas loss in the event of one tank/system failing | Type and capacity of supply to be determined by risk assessment.<br>May not be required when a ring main or other dual supply to a pipeline distribution system is provided |

### Table 3 Liquid cylinder systems

| Primary supply | Secondary supply | Reserve supply<br>(Third source of supply) |
|---|---|---|
| Liquid cylinder manifold system.<br>NB: This is NOT a changeover manifold. All cylinders are on-line simultaneously. | Automatic manifold system.<br>To come on-line in the event of plant failure. | Automatic manifold system.<br>May be sited to support high-dependency areas or whole site<br>OR<br>Locally-based integral valved cylinders with regulators/flow meters attached. |

### Table 4 PSA plant

| Primary supply | Secondary supply | Reserve supply<br>(Third source of supply) |
|---|---|---|
| Multiplex compressors and columns (adsorbers).<br>Subject to design. | Automatic manifold system.<br>To come on-line in the event of plant failure.<br>May be fitted with third party cylinders, or filled from compressor of main plant.<br>Number of cylinders based on ability to provide 4 hours' supply at average use.<br>Locally filled cylinders or gas suppliers' cylinders can be used | Type and capacity of supply to be determined by risk assessment. |

### Table 5 Compressor-driven medical air systems

| Primary supply | Secondary supply | Reserve supply<br>(Third source of supply) |
|---|---|---|
| Duplex compressor system. | Automatic manifold system.<br>To come on-line automatically in the event of plant failure.<br>Number of cylinders based on ability to provide 4 hours' supply at average use. | Automatic manifold system.<br>May be sited to support high-dependency areas or whole site<br>OR<br>Locally-based integral valved cylinders with regulators/flow meters attached. |
| Two compressors of a triplex compressor system. | Third compressor of a triplex system. | Automatic manifold system.<br>To support whole site. |
| Two compressors of a quadruplex system. | Other two compressors of a quadruplex system. | Automatic manifold system.<br>To support whole site. |

### Table 6 Synthetic air plant

| Primary supply | Secondary supply | Reserve supply<br>(Third source of supply) |
|---|---|---|
| Primary oxygen and nitrogen VIE vessels and mixer unit. | Secondary oxygen and nitrogen VIE vessels and mixer unit. | Type and capacity of supply to be determined by risk assessment. |

### Table 7 Combined medical/surgical air plant

| Primary supply | Secondary supply | Reserve supply (Third source of supply) |
|---|---|---|
| Duplex compressor system. | Two automatic manifold systems:<br>• one dedicated to support medical air (MA) system;<br>• one dedicated to support surgical air (SA) system.<br>All to come on-line in the event of plant failure. Number of cylinders on each based on ability to provide 4 hours' supply at average use | Automatic manifold system.<br>May be sited to support high-dependency areas or whole site<br>OR<br>Locally-based integral valved cylinders with regulators/flow meters attached. |
| Two compressors of a triplex compressor system. | Third compressor of a triplex system. | Automatic manifold system.<br>To support whole site. |
| Two compressors of a quadruplex system. | Other two compressors of a quadruplex system. | Automatic manifold system.<br>To support whole site. |

### Table 8 Compressor-driven surgical air systems

| Primary supply | Secondary supply | Reserve supply (Third source of supply) |
|---|---|---|
| Simplex compressor unit. | Automatic manifold system.<br>To come on-line in the event of plant failure.<br>Number of cylinders based on ability to provide 4 hours' supply at average use. | Locally based integral valved cylinders with regulators/flow meters attached. |
| One compressor of a duplex compressor system. | Second compressor of a duplex compressor system. | Automatic manifold system. |

### Table 9 Central medical vacuum systems

| Primary supply | Secondary supply | Reserve supply (Third source of supply) |
|---|---|---|
| Two compressors of a triplex pump system. | Third pump of a triplex system. | Portable suction equipment. |
| Two pumps of a quadruplex system. | Other two pumps of a quadruplex system. | Portable suction equipment. |

**Notes to Tables 1–9:**

General guidance on vacuum systems is contained in Appendix L.

a. Duplex vacuum plant will be classified as a primary source only. A third pump will need to be added to provide a secondary supply to meet the recommendations of this Health Technical Memorandum.

b. For duplex and triplex compressor systems and triplex vacuum pump systems, each compressor/pump will be sized to provide the system's full design flow.

c. For quadruplex systems, each compressor/pump is sized to cope with half the system design flow.

d. For all compressor systems with a design flow greater than 500 L/min, two receivers, each able to be isolated individually, should be installed.

e. All plant is to be connected to the essential electricity supply.

f. For vacuum provision during total electricity supply failure, cylinder- or medical-gas-system-powered vacuum generators can be used.

g. The use of venturi-type vacuum generators is recommended only for emergency use, as these units are generally driven from the medical oxygen system and use large amounts of gas. This can lead to oxygen enrichment and present a potential fire hazard and may result in the emission of pathological material.

h. The manual/secondary manifolds supporting fully automatic manifolds are usually sited with the manifold system. If a risk assessment indicates that this is not in the interests of supply security, they may be sited remotely from the manifold.

In these circumstances, care should be taken to ensure that appropriate backflow protection (or non-return valves) are used to protect the system from failure of either manifold.

j. Manifolds supporting medical air, surgical air and PSA systems should be sited remotely from the compressor systems. Appropriate backflow protection should be provided, as above.

k. Where practicable, a valved by-pass arrangement around compressor and VIE-plant non-return valves should be installed to permit valve replacement without plant shutdown.

m. Fitting non-return valves one pipe size larger will reduce flow resistance, if this is shown to be a critical factor in system design.

n. All sources of supply should be fitted with a test point comprising weatherproof terminal unit and lockable isolating valve.

p. Where medical air is provided by multiple, locally-sited regulators fed from a combined surgical and medical air distribution system, it will be impracticable to connect supporting manifolds at each regulating station. In this situation, extra care should be taken to ensure that the third means of supply is able to support both systems simultaneously. Consideration should be given to additional manifolds sited to support medical air supplies to critical care areas.

## Sizing information for gas supply sources

2.17 Table 10 provides guidance on suggested maximum sizes for gas sources. Final decisions on plant and manifold capacities will depend on both available accommodation and risks to supply security.

**Table 10 Suggested sizes for gas sources**

| Source | Service | Number of cylinders | Cylinder size | Notes |
|---|---|---|---|---|
| **Automatic manifold** | Oxygen | 2 x 10 | J | Used as a stand-alone manifold or support for cryogenic system/PSA plant |
| | Medical air | 2 x 10 | J | Used as a stand-alone manifold or support for compressor plant |
| | Surgical air | 2 x 6 | J | |
| | Oxygen/nitrous oxide mixture | 2 x 8 | G | |
| | Nitrous oxide | 2 x 6 | G | |
| | Carbon dioxide | 2 x 4 | VF | |
| | Helium/oxygen | 2 x 4 | H | |
| | Nitrogen | 2 x 6 | W | |
| **Manual manifold** | Oxygen | 2 x 2 | J | As secondary supply for an automatic manifold system |
| | Medical air | 2 x 2 | J | |
| | Surgical air | 2 x 1 | J | |
| | Oxygen/nitrous oxide mixture | 2 x 2 | G | |
| | Nitrous oxide | 2 x 2 | G | |
| | Carbon dioxide | 2 x 1 | VF | |
| | Helium/oxygen | 2 x 1 | H | |
| | Nitrogen | 2 x 2 | W | |

| Source | Service | Plant size | Notes |
|---|---|---|---|
| **Duplex compressor system** | Medical air | Each compressor sized at full design flow capacity | Receiver water capacity sized at 50% free air delivery (FAD) in 1 minute |
| **Triplex compressor system** | Medical air | Each compressor sized at full design flow capacity | |
| **Quadruplex compressor system** | Medical air | Each compressor sized at half design flow capacity | |
| **Simplex compressor system** | Surgical air | Compressor sized at $^{1}/_{3}$ design flow | Water capacity of receiver sized at 2 x design flow in 1 minute |
| **Duplex compressor system** | Surgical air | Each compressor sized at $^{1}/_{3}$ design flow | |
| **Triplex pump system** | Medical vacuum | Each pump sized at full design flow capacity | Water capacity of reservoir sized at design flow in 1 minute |
| **Quadruplex pump system** | Medical vacuum | Each pump sized at half design flow capacity | |

2.18 Sizing of vacuum-insulated evaporator (VIE) systems, liquid cylinder storage systems, PSA plant and synthetic air plant should be based on historical consumption data and appropriate risk assessments carried out with the medical gas supplier. Allowance should be made for increases in the use of medical gases and changes to the gas demands caused by local developments and strategic issues. For a completely new site, the proposed gas supplier will need to be consulted so that a review of their historical data can be conducted for similar sites. The graph shown in Appendix M will give an approximate indication of expected annual consumption, based on the number of hospital beds. It should be noted that higher consumption could be expected when, for example, high numbers (>20) of continuous positive airway pressure (CPAP) machines are in frequent use (>40 hours per week).

### Notes

a. Automatic manifolds are generally expected to hold a minimum of two days' supply on each bank.

b. Sufficient cylinders for changing one complete bank should be stored in the manifold room for all gases except nitrous oxide/oxygen mixture, for which two complete changes should be stored in the manifold room.

c. Sufficient additional cylinders should be held in the medical gas store to ensure continuous supply for one week.

## Pipeline distribution system design

2.19 The following general information is required to design an MGPS:

a. schedule of provision of terminal units;

b. design flow rates and pressure requirements at each terminal unit;

c. diversified flows for each section of the pipeline system;

d. total flow.

2.20 Guidance on deriving and calculating the above parameters is given in Chapters 3 and 4 of this Part.

2.21 The definition of "departments", which may comprise several wards, treatment rooms etc, should be agreed at the project design stage to avoid confusion.

## Safety

2.22 The safety of an MGPS is dependent on four basic principles:

a. identity;

b. adequacy;

c. continuity;

d. quality of supply.

2.23 Identity is assured by the use of gas-specific connections throughout the pipeline system, including terminal units, connectors etc, and by the adherence to strict testing and commissioning procedures of the system.

2.24 Adequacy of supply depends on an accurate assessment of demands and the selection of plant appropriate to the clinical/medical demands on the system.

2.25 Continuity of supply is achieved by:

- the specification of a system that (with the exception of liquid oxygen systems which may include a secondary vessel) has duplicate components;
- the provision of a third means of supply for all systems except vacuum;
- the provision of alarm systems; and
- connection to the emergency power supply system.

2.26 Surgical air systems are not considered to be life-support systems and therefore duplicate components are not normally required; an emergency/secondary supply is provided.

2.27 Quality of supply is achieved by the use of gases purchased to the appropriate Ph. Eur. requirements or produced by plant performing to specific standards, by the maintenance of cleanliness throughout the installation of the system, and by the implementation of the various testing and commissioning procedures.

## Installation/supply of equipment/maintenance

2.28 The installation of an MGPS should be carried out only by specialist firms registered to BS EN ISO 9001:2000/BS EN ISO 13485:2003 with the scope of registration appropriately defined.

## Modifications

2.29 Special precautions are required when existing installations are to be modified or extended, to ensure that all sections of the pipeline system remaining in use are not contaminated, and that the supply to patients is not compromised. The section to be modified should be physically isolated from the section in use. Closure of isolating valves is insufficient for this purpose. Where area valve service units (AVSUs) and/or line valve assemblies (LVAs) have been installed, blanking spades should be used. This isolation procedure is not required when work is to be carried out on individual terminal units.

2.30 Modification of existing systems may be detrimental to the overall performance of the system. In the case of older systems, there may be insufficient capacity to permit the system to operate safely with the flows typically encountered in use today.

2.31 Any work involving alteration, extension or maintenance work on an existing system should be subject to the permit-to-work procedure (see Part B, Chapter 8).

## Removal of pipework

2.32 Removal and cutting out of redundant medical gas pipelines and equipment can present as great a hazard to patient safety as any other modification. All such removal (including cutting into existing pipelines, and capping off and removal of redundant pipework and equipment) should be carried out by specialist medical gas contractors only. General demolition contractors should not carry out this work.

### Note

Removal of vacuum systems may present additional microbiological hazards and should be undertaken in accordance with routine hygiene practices, that is, covering of open wounds and immediate cleansing and dressing of cuts/scratches received while carrying out the work. Immunisation against certain diseases may be required by the hospital's occupational health department or the employer of tradespeople; therefore, all operatives should ensure that this requirement has been met.

## Validation and verification

2.33 The objective of validation and verification is to ensure that all the necessary safety and performance requirements of the MGPS will be met. Validation and verification procedures will be required for new installations, additions to existing installations and modifications to existing installations. The scope of work will dictate the specific programme required. This is described in Chapter 15.

### Notes

The concept of the existing quality assurance BSI scheme schedule QAS 3720. 1/206/A1 is currently under review. Further guidance will be given when appropriate.

## General fire precautions

### General

2.34 The siting and general structural principles for the design of liquid oxygen storage accommodation are given in Chapter 6, and the requirements for plantrooms and gas manifold rooms in Chapter 14.

2.35 Guidance on cylinder storage and handling is given in Part B.

### Fire detection system

2.36 Smoke or heat detector heads should be installed in the plantrooms, medical gases manifold rooms and (when internal) medical gases cylinder stores in any hospital having a fire detection system in accordance with Health Technical Memorandum 05-03, Part B – 'Firecode: alarm and detection systems'. External stores may also require fire detection systems.

## Electricity supply to medical gas installations

### General

2.37 Electrical installations should be carried out in accordance with the current addition of BS 7671 wiring regulations and associated guidance documents.

2.38 Provision of electrical supply and distribution should take account of guidance issued in Health Technical Memorandum 06-01 – 'Electrical services'.

## Resilience of supply

2.39 Medical gas pipeline systems, associated equipment and alarms are a critical service within a healthcare establishment. Due consideration should be given to ensure the continuity of service under mains power failure conditions.

2.40 Medical gas equipment should be supplied from a dedicated, final sub-circuit which is considered "essential" within the electrical distribution strategy. Alternative means of supply should be considered in the event that internal sub-distribution is compromised.

2.41 In the event of power failure or interruption, all systems should continue to function as they did before the interruption occurred. For example, except for automatic cycling compressors, dryers, pumps etc, the same compressor and dryer (or vacuum pump) set should be on-line, and for manifold systems the same bank should be running.

2.42 All electrical systems, including plant control systems, alarm interfaces etc, should be designed in accordance with electromagnetic compatibility (EMC) directives. For further details, see the "EMC section" within Health Technical Memorandum 06-01.

2.43 It is important that operational managers and designers are fully aware of stand-by electrical supply arrangements and availability and that plans are available to deal with the total loss of electricity under adverse circumstances.

## Electrical installation

2.44 Wiring systems for medical gas installations should be selected in accordance with BS 7671 wiring regulations with particular regard to the environment and risk from mechanical damage. In this regard, PVC-insulated MICS (mineral-insulated copper-sheathed) cable for external/internal locations and heat-rated singles cable in galvanised conduit for plantrooms are considered suitable. For large equipment, fire-rated SWA (steel wire armoured) cable may be appropriate.

2.45 Care should be taken when installing both electrical systems and medical pipeline systems to avoid occasional contact between pipework and electrical cables, conduit or trunking. When physical separation is impractical or contact with extraneous metalwork occurs (for example where the pipeline is carried in metal partitions or where terminal units are mounted on metal bed-head units), the pipeline should be effectively bonded to the metalwork in accordance with BS 7671 wiring regulations.

2.46 The final connection to any equipment (for example alarm panels or control panels) should be made using an unswitched fused connection unit; a double-pole switch should be available to permit work on the equipment.

2.47 Where electrical systems and medical gas pipeline systems are enclosed in a boom, rigid pendant or multi-purpose-type enclosure, care should be taken to ensure that low voltage (LV), extra-low voltage (ELV) and communications and data systems are maintained together but separate from pipeline systems. There should be no access to unprotected live parts within the pendant except by the use of a tool.

## Earthing

2.48 Medical gas pipelines should be bonded together and bonded to the local electrical distribution board in accordance with BS 7671 wiring regulations. The pipelines should not in themselves be used for earthing electrical equipment.

2.49 Flexible pipeline connections, wherever used, should be bonded across the fixed points to ensure earth continuity.

2.50 Where a medical gas outlet or pipeline system is present within a group 2 location as defined by IEE Guidance Note 7 – 'Medical locations', care must be taken to ensure the resistance of the bonding connection is in accordance with the required value.

# 3 Provision of terminal units, and the location of AVSUs, local alarm indicator panels and LVAs

## General

3.1 Terminal unit provision, location of AVSUs, local alarm indicator panels and LVAs are given in Table 11. Medical treatment policy is evolutionary, however, and the project team should review requirements for individual schemes.

## Terminal units

3.2 Terminal units should be mounted in positions that result in the shortest practicable routes for flexible connecting assemblies, between the terminal unit and apparatus. Terminal units may be surface- or flush-mounted. They may also be incorporated with electrical services, nurse call systems, televisions, radio and audio services, in proprietary fittings such as medical supply units, wall panel systems and pendant fittings etc. When they are installed within such fittings, it is essential to maintain the concentricity of the terminal unit bezel with the fascia plate aperture; if the installation is highly eccentric, the bezel will bind on the fascia plate and the terminal unit will not function properly.

3.3 When planning the installation of operating-room pendant fittings, the location of the operating luminaire and other ceiling-mounted devices should be taken into consideration. When the operating room is provided with an ultra-clean ventilation (UCV) system, it may be more practicable (and cost-effective) to have the services (both medical gas and electrical) incorporated as part of the UCV system partial walls. It is particularly advantageous in the case of surgical air systems as rigid pipework can be used, thus avoiding pressure-loss problems that can occur with flexible assemblies used within pendant fittings.

3.4 The following are not permitted:

a. floor-mounted terminal units;

b. vacuum systems in which body or other fluids are drawn through a fixed pipeline connecting a terminal unit or other connector to a remote suction jar.

3.5 All terminal units should conform to BS EN 737-1:1998. Terminal units intended for wall mounting where directly connected equipment such as flow meters are to be used must include a non-swivel device. Terminal units intended for installation with the socket axis vertical, for example in certain types of pendant, or where horizontally mounted but intended for use with indirectly connected equipment by means of a flexible connecting assembly, should also have a non-swivel device because flow meters may be attached. Dimensions of probes are given in BS 5682:2005. It is essential that probes be machined from stainless steel.

3.6 An anaesthetic gas scavenging (AGS) terminal unit should be provided whenever nitrous oxide and anaesthetic agents are available for anaesthetic procedures. In recovery areas, where nitrous oxide is not provided, there is no primary source of anaesthetic gas pollution; thus, no anaesthetic gas scavenging system (AGSS) is required. Guidance on operating departments requires such areas to be mechanically ventilated. Where nitrous oxide mixed with oxygen is provided for analgesic purposes, scavenging is not generally practicable and pollution should therefore be controlled by mechanical ventilation. Details of ventilation requirements are given in Health Building Note 26 (Volume 1) – 'Facilities for surgical procedures'. For dental departments, scavenging is possible by means of nasal masks, and reference should be made to Heath Technical Memorandum 2022 (Supplement 1) – 'Dental compressed air and vacuum systems' (see also Chapter 10).

3.7 The terminal unit (AGS) is specified in ISO 7396. AGSS are covered in Chapter 10.

## Note

Reference should be made to the Department of Health's (1996) 'Advice on the implementation of the Health & Safety Commission's occupational exposure standards for anaesthetic agents'. Further guidance is given in the Health & Safety Executive's (1996) 'Anaesthetic agents: controlling exposure under COSHH'.

3.8 Where respiratory equipment or surgical instruments are serviced, such as in EBME workshops and SSDs, it is normally necessary to install the full range of medical gas terminal units. AGS should be provided as a dedicated system.

3.9 The fixing of terminal units into medical supply systems or to wall surfaces etc should be such that the following forces can be applied:

a. a lateral force of 20 N applied at 100 mm from the surface of the terminal unit without dislodgement or breakage;

b. an axial force of 450 N without dislodgement or breakage.

3.10 Where an array of terminal units is provided at a location, they should be arranged as follows (see Figure 1):

a. for a horizontal array, when viewed from the front, left to right: oxygen, nitrous oxide, nitrous oxide/oxygen mixture (50% v/v), medical air, surgical air, vacuum, anaesthetic gas scavenging, helium/oxygen mixture. If this arrangement is impracticable, a number of rows can be used. For example:

$O_2$, $N_2O$ and/or $N_2O/O_2$

MA, SA, VAC, AGS, $He/O_2$;

b. for a vertical array, with oxygen at the top and in the sequence as for a horizontal array. In many cases a vertical array is impracticable and a more convenient arrangement will comprise a number of rows (see Figure 1);

c. for a circular array, for example where terminal units are installed on the under-surface of a pendant, with the sequence as for a horizontal array, in a clockwise direction when viewed from below. The AGS terminal unit may occupy the centre of such an array.

**Figure 1 Terminal unit mounting order**

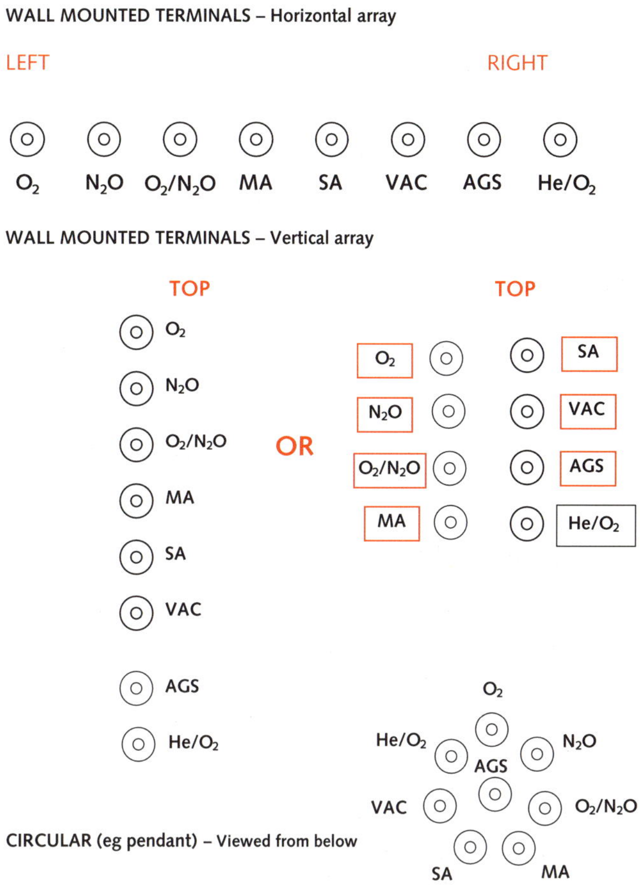


3.11 Oxygen/carbon dioxide mixture systems have been installed, but are no longer covered by this Health Technical Memorandum.

3.12 Helium/oxygen mixtures may be required to be supplied by pipeline in some critical care areas. Systems for these are included in Chapter 11.

3.13 Mounting heights for terminal units should be between 900 mm and 1600 mm above finished floor level (FFL) when installed on walls or similar vertical surfaces – the optimum height for the convenience of users of the medical gas system is 1400 mm (see Figure 2). When terminal units are incorporated within a horizontal bedhead service trunking system, which also provides integrated linear lighting for general room and/or patient reading illumination, it should be of a design that does not compromise the convenience of the medical gas facility.

3.14 When installed in pendants or similar, terminal units should be of a type suitable for mounting within the specified fitting.

3.15 Pressure losses across terminal units should be in accordance with BS EN 737-1:1998. (The standard does not give pressure loss data for surgical air at

350 L/min – but this can be up to 100 kPa when connected via a ceiling NIST (non-interchangeable screw thread) connector and 5 m of hose.)

3.16 Terminal units that are wall mounted should be located as follows (see Figure 2):

a. distance between centres of adjacent horizontal terminal units:

(i) 135 ± 2.5 mm for three or more terminal units;

(ii) 150 ± 2.5 mm for two terminal units only;

b. the distance between the centre of the terminal unit and a potential obstruction on either side (for example when installed in a corner) should be a minimum of 200 mm on either side;

c. care should be taken to ensure that connected medical gas equipment and hoses do not foul other nearby equipment and services during use. Particular attention should be given to terminal unit positioning with respect to worktops, electrical sockets, cupboards, equipment rails, ventilation flaps and door openings. A minimum radial clearance of at least 200 mm from these items is suggested, but this may have to be increased depending on the nature of connected equipment.

### Note

To promote a more "domestic" environment, some in-patient accommodation is provided with terminal units installed in recesses behind covers/decorative panels etc. To accommodate this it is necessary to allow an additional 100 mm on each side of the outermost terminal units and 200 mm from centre to top of recess and 300 mm from centre to bottom of recess. The depth of the recess should be 150 mm. The surface should be clearly marked with suitable legend denoting medical equipment is installed within.

## Terminal units for helium/oxygen mixture

3.17 BS EN 737-1:1998 does not include a terminal unit for helium/oxygen mixture. They will be included in a new edition of BS 5682:1998.

**Figure 2 Terminal unit mounting heights**

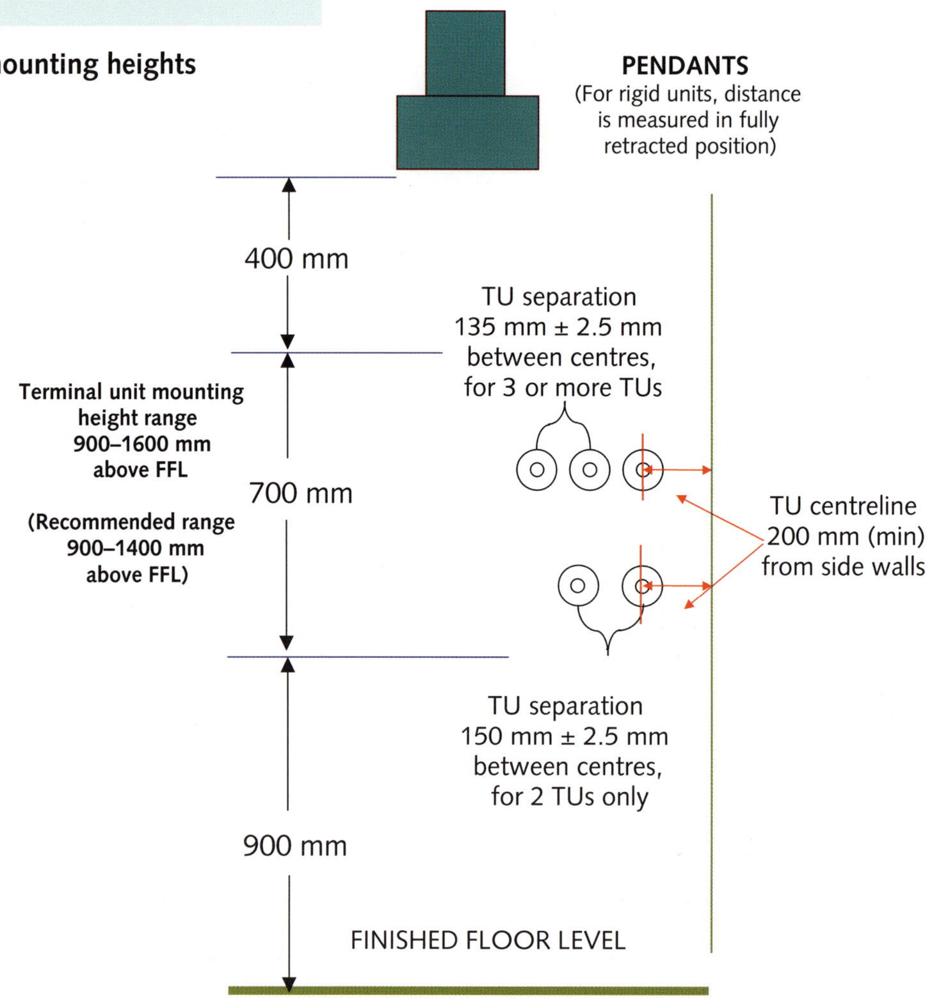

## Nitrogen for surgical tools

3.18 BS EN 739:1998 gives details of connectors for nitrogen for driving tools. The body of the NIST connector should form the wall outlet.

## AVSUs

3.19 AVSUs should be mounted at a convenient height between 1 m and 1.8 m such that they can be operated comfortably by staff without their needing to stoop or overreach (see Figure 3). The order of the location of individual valves in an array should follow that for terminal units, for example: $O_2$, $N_2O$ and/or $N_2O/O_2$, MA, SA, VAC, $He/O_2$. If the array exceeds 1 m in height from top to bottom, it may be preferable to arrange them in two columns. Care must be taken to ensure that AVSUs cannot be obscured by opening doors etc. Details of the design of AVSUs are given in Chapter 13.

### Note

The minimum height of 1 m is the optimum. In critical care areas where dual circuits are installed, it may be necessary to reduce this to 800 mm to avoid an excessive number of columns of AVSUs.

## Local alarm indicator panels

3.20 The placing of local alarm indicators should be such that they are readily visible by staff; notices, partitioning, screens etc should not obscure them. The mounting height should be such that in the event of an audible alarm sounding, staff can activate the "mute" switch without overreaching, and be a maximum 1.8 m above finished floor level (see Figure 3).

## LVAs

3.21 LVAs should be installed at branches from risers, branches from main runs, and where pipelines pass into or out of a building. Details of the design of LVAs are given in Chapter 13.

**Figure 3 AVSU and local alarm panel mounting heights**

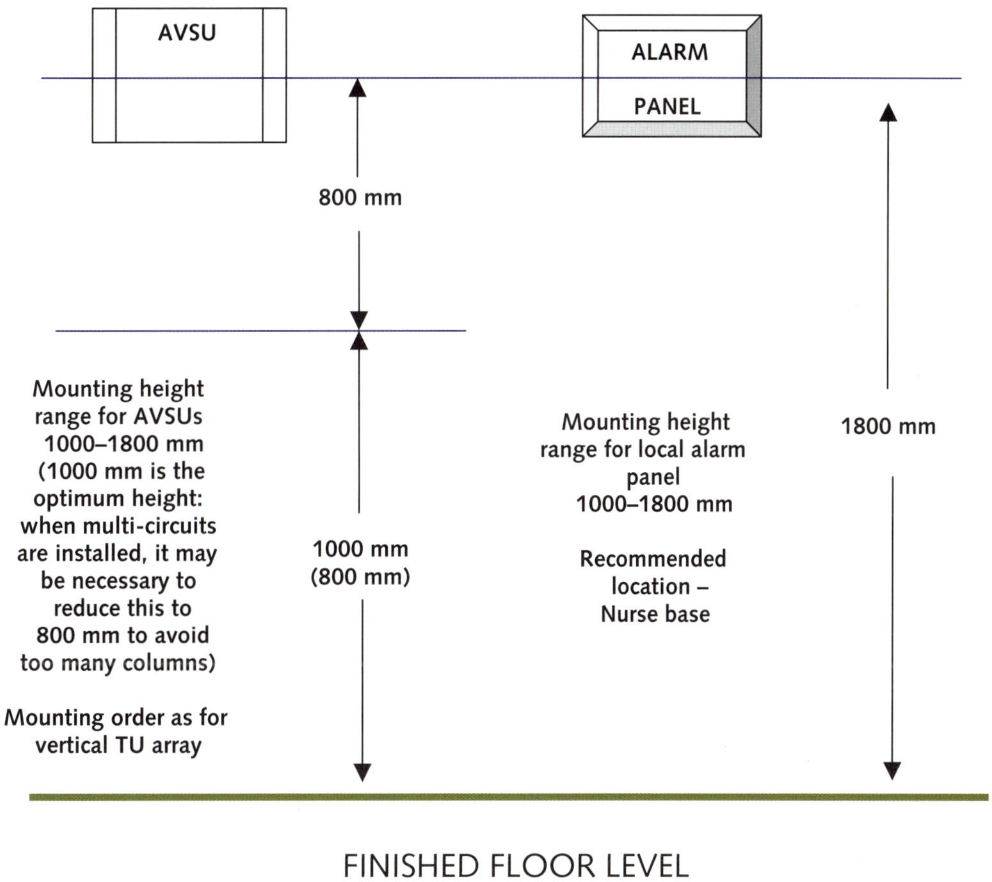

## Specific labelling requirements

3.22 All AVSUs should be labelled to identify the individual rooms, sets of terminal units etc controlled. They should be provided with flow direction arrows.

3.23 In critical care areas where dual circuits and/or subdivision of circuitry occur, terminal units require to be identified as associated with the specific AVSU. Correspondingly, AVSUs should be similarly labelled to identify the terminal units controlled.

**Table 11 Provision of terminal units, AVSUs and local alarms**

| Department | $O_2$ | $N_2O$ | $N_2O/O_2$ | MA4 | SA7 | VAC | AGSS | $He/O_2$ | AVSU | Alarm |
|---|---|---|---|---|---|---|---|---|---|---|
| **Accident and Emergency** | | | | | | | | | 1 set (1) | 1 set hp/lp (9) |
| Resuscitation room, per trolley space | 2 | 2 | – | 2 | – | 2 | 2 | – | 2 sets* | |
| Note: One set either side of the trolley space, if installed in fixed location, eg trunking; or both sets in an articulated supply pendant that can be positioned either side of the bed space. | | | | | | | | | | |
| Major treatment/plaster room per trolley space | 1 | 1 | 1p | 1 | 1p | 1 | 1 | – | 1 set/8 TUs | |
| Post-anaesthesia recovery per trolley space | 2 | – | – | 2 | – | 2 | – | – | 2 sets* | |
| Note: One set either side of the trolley space, if installed in fixed location, eg trunking; or both sets in an articulated supply pendant that can be positioned either side of the bed space. | | | | | | | | | | |
| Treatment room/cubicle | 1 | – | – | – | – | 1 | – | – | 1 set/8 TUs | |
| **Operating department** | | | | | | | | | 1 set (1) | |
| Anaesthetic rooms (all) | 1 | 1 | – | 1 | – | 1 | 1 | – | | |
| Operating room, orthopaedic: | | | | | | | | | | |
| For anaesthetist | 2 | 1 | – | 2 | – | 2 | 1 | – | 1 set per suite (2)(3) | 1 set per suite hp/lp (10) |
| For surgeon | – | – | – | – | 4 | 2 | – | – | – | – |
| Note: Orthopaedic surgery is normally performed in operating rooms provided with ultra-clean systems. Such systems are much more effective in terms of airflow when provided with partial walls. These walls may be effectively used to include terminal units that can be supplied by rigid pipework. Such installations do not suffer from excessive pressure loss when surgical air is required at high flows. | | | | | | | | | | |
| Operating room, neurosurgery | | | | | | | | | | |
| Anaesthetist | 2 | 1 | – | 2 | – | 2 | 1 | – | 1 set per suite (2)(3) | 1 set per suite hp/lp (10) |
| Surgeon | – | – | – | – | 2 | 2 | – | – | | |
| Note: If multi-purpose pendants are used, there may be some loss of performance of surgical tools because of bore restrictions and convolution of the flexible connecting assemblies at the articulated joints. | | | | | | | | | | |

| Department | $O_2$ | $N_2O$ | $N_2O/O_2$ | MA4 | SA7 | VAC | AGSS | $He/O_2$ | AVSU | Alarm |
|---|---|---|---|---|---|---|---|---|---|---|
| Operating room, general surgery etc | | | | | | | | | | |
| Anaesthetist/surgeon (see (14) regarding $CO_2$) | 2 | 2 | – | 2 | 2p | 2 | 2 | – | 1 set per suite (2)(3) | 1 set per suite hp/lp (10) |
| Note: Terminal units installed in separate pendants: p = project team option where some orthopaedic overspill surgery may be performed. | | | | | | | | | | |
| Post-anaesthesia recovery, per bed space | 2 | – | – | 2 | – | 2 | – | – | 2 sets* (5) | 1 alarm for both sets of AVSUs (11) |
| Note: One set either side of the bed space, if installed in fixed location, eg trunking; or both sets in an articulated supply pendant that can be positioned either side of the bed space. | | | | | | | | | | |
| Equipment service room per work space | 1 | 1 | – | 1 | 1 | 1 | 1 | – | 1 set | 1 set hp/lp (12) |
| **Maternity department** | | | | | | | | | 1 set (1) | 1 set (9) |
| LDRP room (normal/abnormal) | | | | | | | | | 1 set per 6–8 rooms (5) | |
| Mother | 1 | – | 1 | – | – | 2 | – | – | | |
| Baby (per cot space) (allow for 2 cots only) | 1 | – | – | 1 | – | 1 | – | – | | |
| Operating suite: | | | | | | | | | | |
| Anaesthetist | 1 | 1 | – | 1 | – | 1 | 1 | – | 1 set | 1 set hp/lp (10) |
| Obstetrician | – | – | – | – | – | 2 | – | – | | |
| Paediatrician (per cot space) (allow for 2 cots only) | 1 | – | – | 1 | – | 1 | – | – | | |
| Post-anaesthesia recovery (per bed space) | 1 | – | – | 1 | – | 1 | – | – | 1 set | 1 set (11) |
| Equipment service room** per work space | 1 | 1 | 1 | 1 | – | 1 | 1 | – | 1 set | 1 set (12) |
| Neonatal unit, per cot space | 2 | – | – | 2 | – | 2 | – | – | 2 sets* | 1 for both sets of AVSUs (11) |
| Note: One set either side of the bed space, if installed in fixed location, eg trunking; or both sets in an articulated supply pendant that can be positioned either side of the bed space. | | | | | | | | | | |
| Equipment service room** per work space | 1 | – | – | 1 | – | 1 | – | – | 1 set | 1 set (12) |
| In-patient accommodation: | | | | | | | | | | |
| Single bed room | 1 | – | – | – | – | 1 | – | – | 1 set for ward unit | 1 set (11) |
| Multi-room, per bed space | 1 | – | – | – | – | 1 | – | – | | |
| Nursery, per cot space | 1 | – | – | – | – | 1 | – | – | | |
| Provision for 2 cots only irrespective of number of cot spaces | | | | | | | | | | |

| Department | $O_2$ | $N_2O$ | $N_2O/O_2$ | MA4 | SA7 | VAC | AGSS | $He/O_2$ | AVSU | Alarm |
|---|---|---|---|---|---|---|---|---|---|---|
| **Diagnostics departments** | | | | | | | | | 1 set (1) | 1 set (9) |
| Special procedures room | 1 | 1 | – | 1 | – | 1 | 1 | – | 1 set | 1 set (10) |
| Anaesthetic room | 1 | 1 | – | 1 | – | 1 | 1 | – | 1 set | |
| Holding and recovery | 1 | – | – | 1p | – | 1 | – | – | 1 set | |
| Ultrasound | 1 | – | – | – | – | 1 | – | – | 1 set | |
| Fluoroscopy | 1 | – | – | – | – | 1 | – | – | 1 set | |
| Urography | 1 | – | – | – | – | 1 | – | – | 1 set | |
| Tomography | 1 | – | – | – | – | 1 | – | – | 1 set | |
| Magnetic resonance imaging (MRI) suite | 1 | 1 | – | 1 | – | 1 | 1 | – | 1 set | 1 set (10) |
| CAT room | 1 | 1 | – | 1 | – | 1 | 1 | – | 1 set | 1 set (10) |
| Angiography | 1 | 1 | – | 1 | – | 1 | 1 | – | 1 set | 1 set (10) |
| Endoscopy | 1 | 1 | – | 1 | – | 1 | 1 | – | 1 set | 1 set (10) |
| Lineac bunkers | 1 | 1 | – | 1 | – | 1 | 1 | – | 1 set | 1 set (10) |
| General purpose rooms | 1 | – | – | – | – | 1 | – | – | | |
| **In-patient accommodation** | | | | | | | | | 1 set for the ward unit (1) | |
| Single-bed room | 1<br>1 | –<br>– | –<br>– | 1<br>1 | –<br>– | 1<br>1 | –<br>– | –<br>– | | 1 set (11) |
| Multi-bed room, per bed space | 1 | – | – | 1 | – | 1 | – | – | | |
| Treatment room (Appropriate for adult acute, children and the elderly) | 1 | – | – | 1 | – | 1 | – | – | | |
| **Renal** | | | | | | | | | | |
| Per dialysis station | 1 | – | – | 1 | – | 1 | – | – | | |
| Per bed space | 1 | – | – | 1 | – | 1 | – | – | | |
| **Critical care area** | | | | | | | | | 1 set (1) | 1 set (11) |
| Per bed space | 4 | 2p | 2p | 4 | | 4 | 2p | 2p | 2 sets* (4) | |
| Note: One set either side of the bed space, if installed in fixed location, eg trunking; or both sets in an articulated supply pendant that can be positioned either side of the bed space. | | | | | | | | | | |
| Equipment service room, per work space | 1 | 1p | 1p | 1 | | 1 | 1p | | 1 set | 1 set (12) |
| Coronary care unit (CCU) | | | | | | | | | | |
| Per bed space | 4 | | | 4 | | 4 | | | 2 sets* (4)(6)(7) | 1 for both sets of AVSUs (11) |

| Department | $O_2$ | $N_2O$ | $N_2O/O_2$ | MA4 | SA7 | VAC | AGSS | $He/O_2$ | AVSU | Alarm |
|---|---|---|---|---|---|---|---|---|---|---|
| High dependency unit (HDU) | | | | | | | | | | |
| Per bed space | 4 | | | 4 | | 4 | | | 2 sets* (4)(6)(7) | 1 for both sets of AVSUs (11) |
| Burns unit | 2 | 2p | 2p | 2 | | 2 | 2p | | 2 sets* (4)(6)(7) | 1 for both sets of AVSUs (11) |
| **Adult mental illness accommodation** | | | | | | | | | 1 set (1) | |
| Electro-convulsive therapy (ECT) room | 1 | 1 | | 1 | | 1 | 1 | | 1 set (2) | 1 set hp/lp (10) |
| Post-anaesthesia recovery, per bed space | 1 | | | 1 | | 1 | | | 1 set (1) | 1 set (11) |
| **Adult acute day care accommodation** | | | | | | | | | 1 set | |
| Treatment room: | | | | | | | | | | |
| Anaesthetist | 1 | 1p | | 1p | | 1 | 1p | | 1 set | 1 set hp/lp |
| Surgeon | | | | | | 2 | | | | if (p) (10) |
| Post anaesthesia recovery, per bed space | 1 | | | 1p | | 1 | | | 1 set | 1 set (11) |
| **Day patient accommodation** | | | | | | | | | 1 set (1) | |
| Single bed room | 1 | | | | | 1 | | | 1 set (1)(13) | 1 set hp/lp (11) |
| Multi-bed room, per bed space | 1 | | | | | 1 | | | | |
| Treatment room | 1 | | | 1p | | 1 | | | | |
| Endoscopy room | 1 | 1p | | 1p | | 1 | 1p | | 1 set (2)(13) | 1 set (10) |
| **Fracture clinic** | | | | | | | | | | |
| Plaster room | 1 | 1 | | 1p | 1p | 1 | 1 | | 1 set (1) | 1 set (9) |
| **Oral surgery, orthodontic department** | | | | | | | | | 1 set (1) | 1 set hp/lp if (p) (9) |
| Consulting/treatment room, type 1 | 1 | 1p | | 1 | Dental air will be required | † | 1p | | 1 set/4–6 rooms | |
| Consulting/treatment room, types 2 and 3 | 1 | | | 1 | | † | | | 1 set/4–6 rooms | |
| Recovery room, per recovery position | 1 | | | | 1 | 1 | | | 1 set | |
| Appliance laboratory, per workstation | 1 | 1p | | 1 | 1 | 1 | 1p | | 1 set | 1 set hp/lp (11) |
| **Out-patient department** | | | | | | | | | | |
| Treatment room/cubicles | 1 | 1p | | 1p | | 1 | 1p | | 1 set (1)(8) | 1 set hp/lp if (p) (9) |

| | | | | | | | | | | |
|---|---|---|---|---|---|---|---|---|---|---|
| **Sterile services department** | | | | | | | | | | |
| Wash room | | | | 1 | 1 | | | | | |
| Inspection, assembly and packing (IAP) room | 1 | | | 1 | 1 | | | 1 | 1 set (1) | 1 set (9) |
| Other departments (as required by project team) | | | | | | | | | 1 set (1) | 1 set located (9–12) as appropriate |

**Notes:**

* Dual circuits.

** Where the delivery and neonatal units are in close proximity, the equipment service room can be shared.

† Dental vacuum only.

p = Project team option.

hp/lp = high-pressure and low-pressure alarms for oxygen, medical air and nitrous oxide when installed together. All other local alarms, low pressure only.

(1) Departmental AVSUs installed on the hospital street side of fire compartment doors.

(2) Installed immediately outside the room.

(3) Where air is used to control movable pendant fittings, it should be taken from the 7 bar surgical air system.

(4) In addition to the dual circuits, additional AVSUs will be required to sub-divide the number of terminal units controlled. This subdivision should be based on the layout of the accommodation; for example, if the recovery area is divided into a number of separate room/areas, each would have a separate sub-set (see Figures 4 and 5).

(5) This is intended to provide some flexibility and the exact number will depend on the total number of rooms within the department.

(6) If a high-dependency unit is included within general in-patient accommodation, a separate set of AVSUs should be provided for the unit. In addition to the departmental valves or the ward as a whole, an additional set will be required to control the single-bed, multi-bed and treatment rooms.

(7) Department AVSUs may be required if the units are large and separate from, for example, the critical care area.

(8) Additional AVSUs may be required in a large unit: the aim should be to have about 8–12 rooms controlled by a set of valves – discretion is required to arrive at the logical number.

(9) Installed in reception area.

(10) Installed in the operating room in the "main panel" or within the room, or an ante-room, eg control room of an MRI device.

(11) Installed at the main staff base (nurses' station).

(12) Installed in the room space with the AVSUs.

(13) Separate AVSUs will be required if endoscopy room is included.

(14) Carbon dioxide is used for insufflation during some surgical procedures. A pipeline installation is a project team option and is covered in Chapter 11. Two NIST connector bodies units should be installed.

**General:**

Normally, departmental AVSUs would be installed at the hospital street side of the entrance doors to a department and would reflect the method of horizontal evacuation in the event of an emergency. In some large departments, for example an operating department, the clean-service corridor is likely to cross one or more fire compartment walls. Additional AVSUs may therefore be required to reflect the evacuation route.

If a department includes one or more floors, a set of AVSUs should be provided for each floor, which will act as emergency overall fire valves.

AVSUs for zones within critical care areas should be located where they can be seen by staff – not necessarily at the staff base.

Local alarms within critical care areas should be provided for the individual space; that is, if a critical care area of, say, 18 beds is sub-divided into three separate six-bed wards, there should be one alarm only for each space (not one for each of the dual circuits).

**Figure 4 Larger critical care area with isolation room and twin four-bed bays**

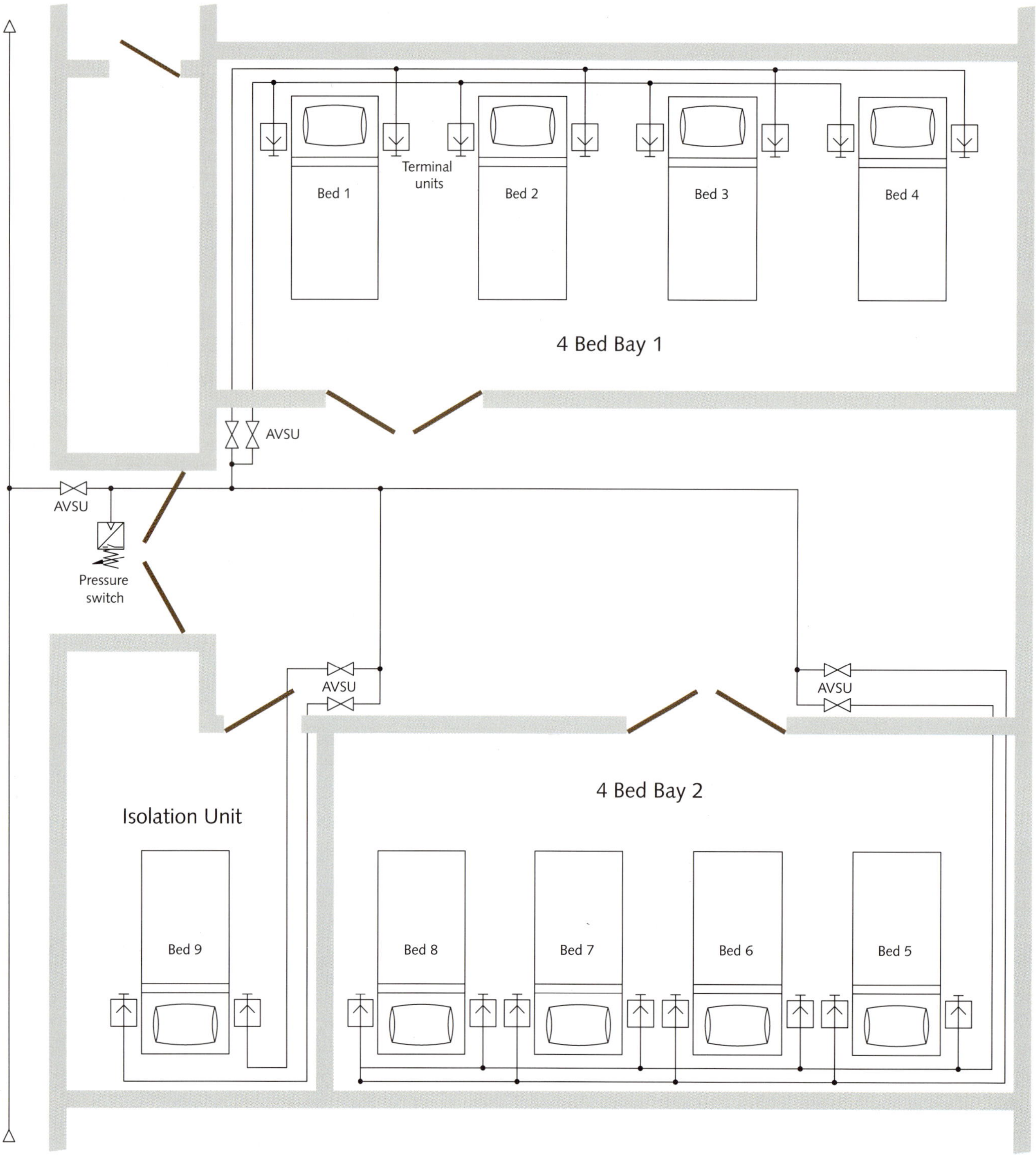

## Figure 5 Smaller critical care area with isolation room and five-bed bay

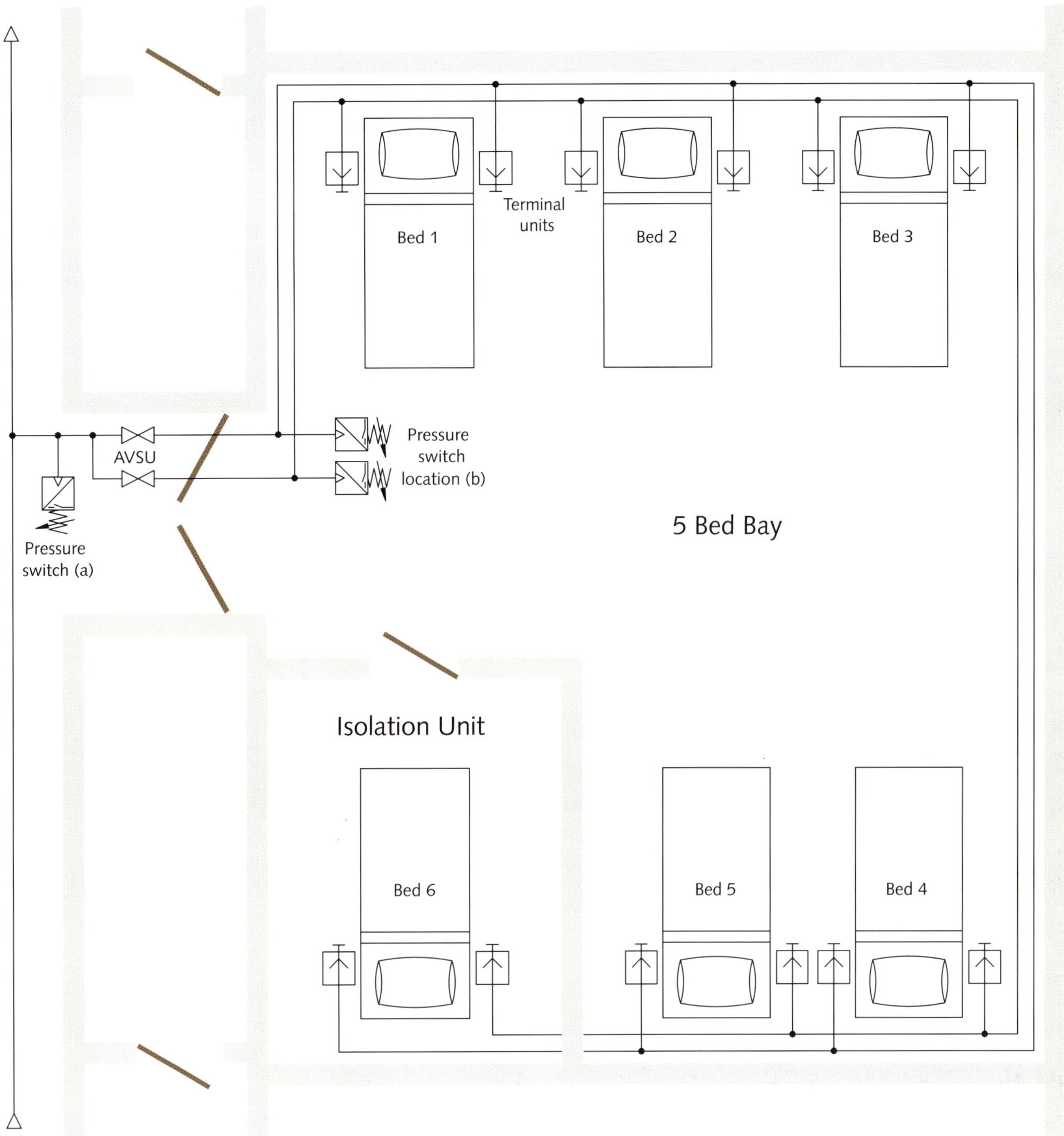


Note:
Pressure switch location is decided as follows:

(a) upstream of unit AVSUs assumes a departmental AVSU is upstream of pressure switch;
(b) downstream of unit AVSUs will require two pressure switches connected in parallel, one for each circuit.

# 4 Gas flow

## General

4.1 Various layouts of an MGPS are shown throughout this document, and each will need to be designed to take into account the anticipated design flow. Appendix N provides a conversion table for various units of measurement that may be encountered.

4.2 There are several aspects of gas flow to consider when designing the pipeline distribution system:

a. the test flow that is required at each terminal unit for test purposes (this flow is essentially to establish that the terminal unit functions correctly and that there are no obstructions; see Table 12);

b. the typical flow required at each terminal (this is the maximum flow likely to be required at any time in clinical use; see Table 12);

c. the likely numbers of terminal units in use at any time;

d. the flow required in each sub-branch of the distribution, for example from the terminal unit or a number of terminal units (for example four in a four-bed ward) to the pipeline in the false ceiling of the ward corridor;

e. the total flow to the ward/department, that is, the sum of the diversified flows in each sub-branch;

f. the flow in the main branches/risers, that is, the summation of all diversified flows;

g. the flow required at the plant. In most cases this will be the flow in (f) above except in the case of vacuum that is not used continuously.

4.3 The pipeline system should be designed so that the flows given in Table 12 can be achieved at each terminal unit: the flows are expressed in free air. Diversified flows are used for the purposes of pipe size selection.

4.4 The designer should always ensure that due account is taken of the stated use of a particular department.

4.5 There is a limited range of pipe sizes, and where there is any doubt about flow requirements, a larger pipe size should be selected.

### Note

When calculating diversified flows, it is the number of bed spaces, treatment spaces or rooms in which the clinical procedure is being performed that is used; this is not the individual number of terminal units since, in many cases, more than one is installed. For example, a bed position in a critical care area may have four or more oxygen terminal units.

4.6 The overall pipeline design should be based on a 5% pressure drop from the plant/source of supply to that measured at the terminal unit outlet at the specified test flows.

## Terminal unit flows

4.7 At the design stage, the project team should define the individual room/space requirements. Departments usually comprise several ward units, treatment rooms and other spaces. In order to avoid confusion, the nomenclature for each clinical space should be clearly defined so that the appropriate gas flow requirements can be established at the commencement of the design stage.

## Pipeline flows

4.8 Precise prediction of pipeline flow is not possible, but there are guidelines that can be used. and these have been shown to be adequate in practice.

4.9 For vacuum systems, the minimum vacuum should not fall below 300 mm Hg at the front of each terminal unit at a design flow of 40 L/min.

4.10 The design of the pipework system is based on the diversified flows and the permissible pressure loss from the source of supply to, and including, the terminal unit pressure loss. The pipe sizes should be selected to ensure that the pressure loss is below 5%

## Table 12 Gas flow – flows required at terminal units

| Service | Location | Nominal pressure (kPa) | Design flow (L/min) | Typical flow required (L/min) | Test flow (L/min) |
|---|---|---|---|---|---|
| Oxygen | Operating rooms and rooms in which $N_2O$ is provided for anaesthetic purposes | 400 | 100[(1)] | 20 | 100 |
| | All other areas | 400 | 10 | 6 | 40 |
| Nitrous oxide | All areas | 400 | 15 | 6 | 40 |
| Nitrous oxide/ oxygen mixture | LDRP (labour, delivery, recovery, post-partum) rooms | 310[(2)] | 275 | 20 | 275 |
| | All other areas | 400 | 20 | 15 | 40 |
| Medical air 400 kPa | Operating rooms | 400 | 40[(3)] | 40 | 80 |
| | Critical care areas, neonatal, high dependency units | 400 | 80[(3)] | 80 | 80 |
| | Other areas | 400 | 20 | 10[(3)] | 80 |
| Surgical air/ nitrogen | Orthopaedic and neurosurgical operating rooms | 700 | 350[(4)] | 350 | 350 |
| Vacuum | All areas | 40 (300 mm Hg below atmospheric pressure) | 40 | 40 maximum, further diversities apply | 40 |
| Helium/oxygen mixture | Critical care areas | 400 | 100 | 40 | 80 |

Notes:

1. During oxygen flush in operating and anaesthetic rooms.

2. Minimum pressure at 275 L/min.

3. These flows are for certain types of gas-driven ventilator under specific operating conditions, and nebulisers etc.

4. Surgical air is also used as a power source for tourniquets.

## Figure 6 Typical pressures in medical air/oxygen/nitrous oxide/nitrous oxide-oxygen mixture systems under design flow conditions

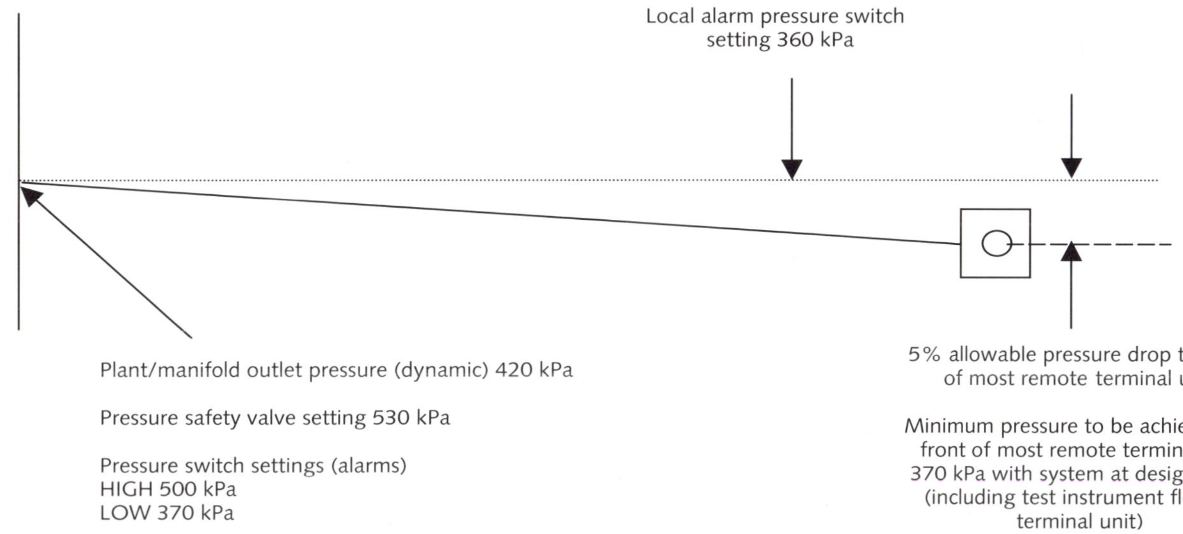

**Figure 7 Typical pressures in a vacuum system under design flow conditions**

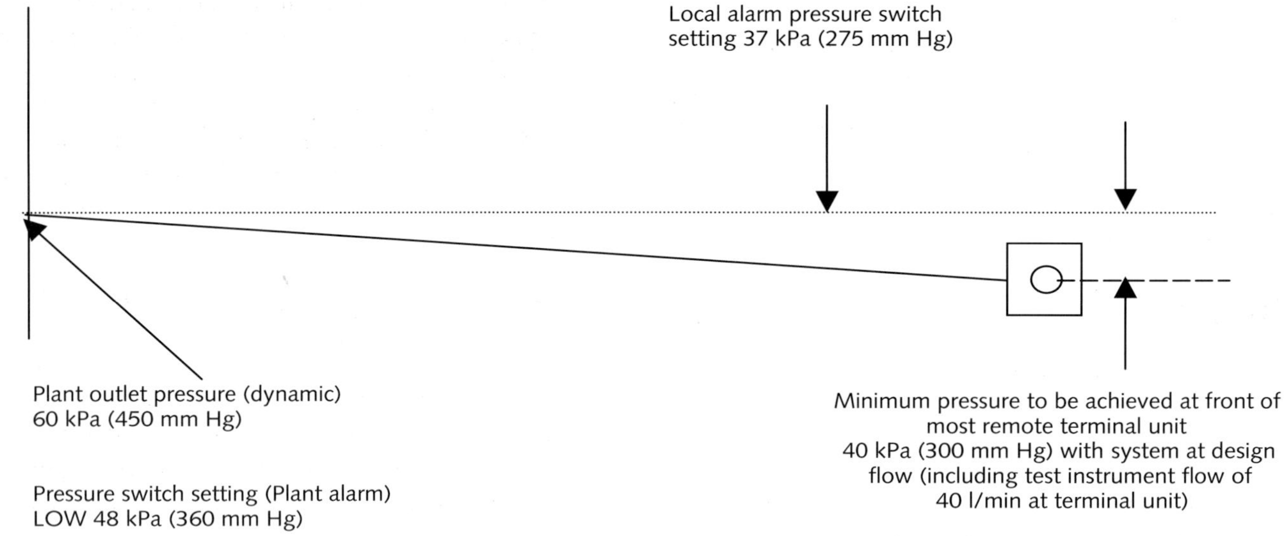


**Figure 8 Typical pressures in a single pressure reduction surgical air system under design flow conditions**

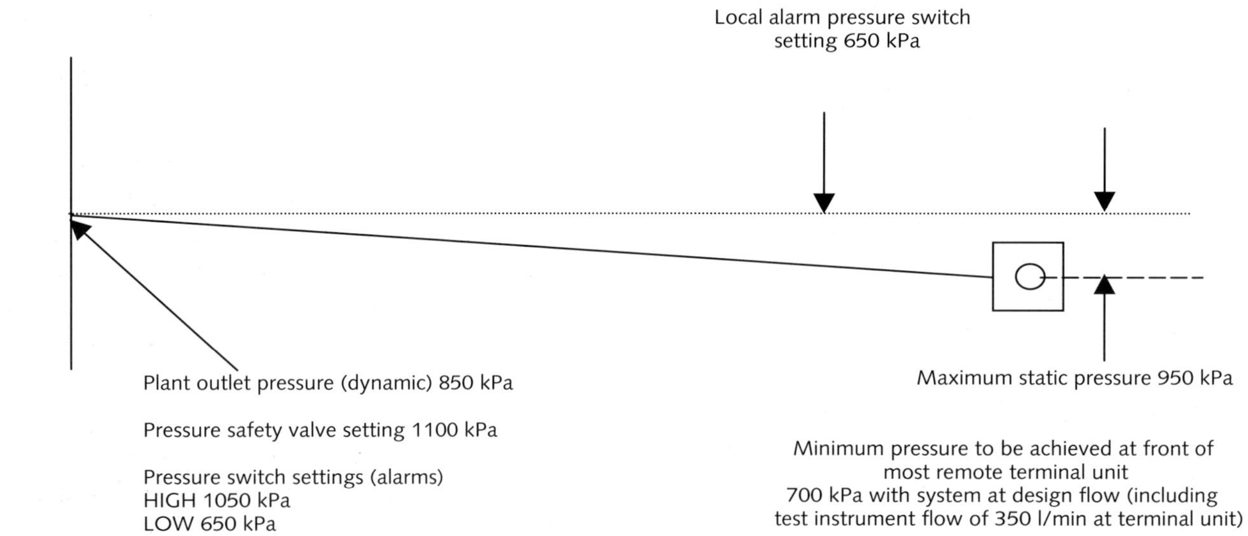


**Figure 9 Typical pressures in a double pressure reduction surgical air system under design flow conditions**

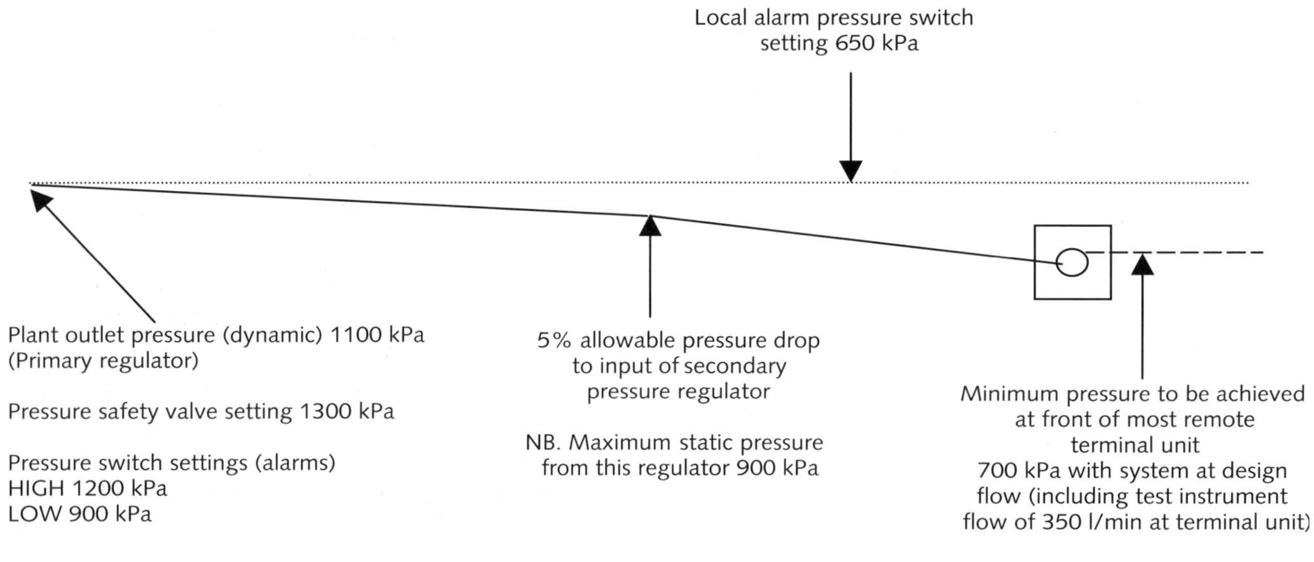

of the nominal pipeline pressure (see Figures 6–9, and Appendix G).

4.11 Pressure requirements for surgical air are based on the requirement that the minimum pressure should be 700 kPa at the terminal unit at a flow of 350 L/min.

4.12 Details of pressure requirements for all systems are described in paragraphs 4.43–4.50.

## Oxygen

### In-patient accommodation

4.13 Oxygen is used at a typical flow of 5–6 L/min. Each terminal unit should, however, be capable of passing 10 L/min (at standard temperature and pressure (STP)) at a supply pressure of 400 kPa (nominal) as shown in Table 12, in case nebulisers or other respiratory equipment are used. Table 13 contains the formula for arriving at diversified flows.

4.14 For a 28-bed ward unit comprising single and four-bed rooms and a treatment room, the diversified flow is calculated on the assumption that one bed space requires 10 L/min, and one in four of the remainder require 6 L/min. For the purpose of pipe size selection, the diversified flow at entry to the ward is taken as 50 L/min (strictly 50.5 L/min); it is assumed that a patient will use oxygen in a ward or in the treatment room but not both.

4.15 When selecting the size of a sub-branch serving, for example, a four-bed ward, the flow would be taken to be 28 L/min as all four in-patients could be using oxygen; for larger wards no additional flow is added until the formula in Table 13 comes into play.

4.16 A department may comprise several ward units as above. The diversified flow for each department $Q_d$ is based on $Q_w$ for the first ward unit, plus 50% of the flow for the remaining ward units. For the purposes of this calculation, the first ward unit is taken as the largest within the department.

4.17 If one ward unit is significantly larger than the others, the flows from the ward units should be averaged to obtain a more realistic value.

### Operating departments

4.18 The diversified flow for operating departments is based on 100 L/min required for the oxygen flush. Therefore each oxygen terminal unit in the operating room and anaesthetic room should be able to pass 100 L/min. It is unlikely that an oxygen flush will be administered simultaneously in several operating rooms. The diversified flow $Q$ is based on 100 L/min for the first operating room and 10 L/min for the remainder. To obtain the flow to each operating suite, add together the flows for the operating and anaesthetic room, that is, 110 L/min.

4.19 For anaesthetic rooms, each terminal unit should be capable of passing 100 L/min (it may be necessary to use oxygen "flush"), but the actual flow likely to be used is 6 L/min or less. As it is unlikely that a patient would be anaesthetised at the same time that a patient in the associated operating room was continuing to be treated under an anaesthetic (and because the duration of induction is short), no additional flow is included.

4.20 In recovery, it is possible that all bed spaces may be in use simultaneously; hence, no diversity is used.

### Critical care, coronary care and high-dependency units

4.21 The flow for these units assumes that, although all bed spaces may be occupied, three-quarters of these will require the use of oxygen. Each terminal unit should be capable of delivering 10 L/min. The diversified flow is calculated assuming 10 L/min for the first bed space and 6 L/min for three-quarters of the remainder.

4.22 Oxygen should not be used as the driving gas for gas-powered ventilators if they are capable of being powered by medical air. The minimum flow that has been shown to be adequate to drive current types of ventilator is 80 L/min at 360 kPa. For test purposes the minimum pressure is 370 kPa.

4.23 If oxygen has to be used to power ventilators and/or ventilators are operating in CPAP mode, the high flows that may be encountered should be taken into account both when designing the pipeline and when sizing the supply vessel. These ventilators use exceptional amounts of oxygen, particularly if adjusted incorrectly. If incorrectly set, they can use in excess of 120 L/min, but their therapeutic benefit will be effective at lower flows. To allow for some flexibility, and additional capacity, a diversified flow of 75 L/min for 75% of beds has been included. If significant numbers of beds are required to treat patients using CPAP ventilation, consideration should be given to running a separate pipeline from the source of

## Table 13 Oxygen: design and diversified flows

| Department | Design flow for each terminal unit (L/min) | Diversified flow $Q$ (L/min) |
|---|---|---|
| **In-patient accommodation (ward units):** | | |
| Single 4-bed rooms and treatment room | 10 | $Q_w = 10 + [(n - 1)6/4]$ |
| Ward block/department | 10 | $Q_d = Q_w[1 + (nW - 1)/2]$ |
| **Accident & emergency:** | | |
| Resuscitation room, per trolley space | 100 | $Q = 100 + [(n - 1)6/4]$ |
| Major treatment/plaster room, per trolley space | 10 | $Q = 10 + [(n - 1)6/4]$ |
| Post-anaesthesia recovery, per trolley space | 10 | $Q = 10 + [(n - 1)6/8]$ |
| Treatment room/cubicle | 10 | $Q = 10 + [(n - 1)6/10]$ |
| **Operating:** | | |
| Anaesthetic rooms | 100 | $Q$ = no addition made |
| Operating rooms | 100 | $Q = 100 + (nT - 1)10$ |
| Post-anaesthesia recovery | | $Q = 10 + (n - 1)6$ |
| **Maternity:** | | |
| LDRP rooms: | | |
| Mother | 10 | $Q = 10 + [(n - 1)6/4]$ |
| Baby | 10 | $Q = 10 + [(n - 1)3/2]$ |
| Operating suites: | | |
| Anaesthetist | 100 | $Q = 100 + (nS - 1)6$ |
| Paediatrician | 10 | $Q = 10 + (n - 1)3$ |
| Post-anaesthesia recovery | 10 | $Q = 10 + [(n - 1)3/4]$ |
| In-patient accommodation: | | |
| Single/multi-bed wards | 10 | $Q = 10 + [(n - 1)6/6]$ |
| Nursery, per cot space | 10 | $Q = 10 + [(n - 1)3/2]$ |
| Special care baby unit | 10 | $Q = 10 + (n - 1)6$ |
| **Radiological:** All anaesthetic and procedures rooms | 100 | $Q = 10 + [(n - 1)6/3]$ |
| **Critical care areas** | 10 | $Q = 10 + [(n - 1)6]3/4$ |
| **Coronary care unit (CCU)** | 10 | $Q = 10 + [(n - 1)6]3/4$ |
| **High-dependency unit (HDU)** | 10 | $Q = 10 + [(n - 1)6]3/4$ |
| **Renal** | 10 | $Q = 10 + [(n - 1)6/4]$ |
| **CPAP ventilation** | 75 | $Q = 75n \times 75\%$ |
| **Adult mental illness accommodation:** | | |
| Electro-convulsive therapy (ECT) room | 10 | $Q = 10 + [(n - 1)6/4]$ |
| Post-anaesthesia, per bed space | 10 | $Q = 10 + [(n - 1)6/4]$ |
| **Adult acute day care accommodation:** | | |
| Treatment rooms | 10 | $Q = 10 + [(n - 1)6/4]$ |
| Post-anaesthesia recovery per bed space | 10 | $Q = 10 + [(n - 1)6/4]$ |
| **Day patient accommodation** (as "In-patient accommodation") | | As "In-patient accommodation" |
| **Oral surgery/orthodontic:** | | |
| Consulting rooms, type 1 | 10 | $Q = 10 + [(n - 1)6/2]$ |
| Consulting rooms, types 2 & 3 | 10 | $Q = 10 + [(n - 1)6/3]$ |
| Recovery room, per bed space | 10 | $Q = 10 + [(n - 1)6/6]$ |
| **Out-patient:** | | |
| Treatment rooms | 10 | $Q = 10 + [(n - 1)6/4]$ |
| **Equipment service rooms, sterile services etc** | 100 | Residual capacity will be adequate without an additional allowance |

**Legend for Tables 13–21:**

$Q$ = diversified flow for the department;

$Q_w$ = diversified flow for the ward;

$Q_d$ = diversified flow for the department (comprising two or more wards);

$n$ = number of beds, treatment spaces or single rooms in which the clinical procedure is being performed, not the individual numbers of terminal units where, in some cases, more than one is installed;

$nS$ = number of operating suites within the department (anaesthetic room and operating room).

$nW$ = number of wards

$nT$ = number of theatres

---

supply. Care should be taken when calculating air exchange rates in wards/rooms in which large numbers of CPAP machines may be in use simultaneously and where failure of mechanical ventilation could result in raised ambient oxygen concentrations. Consideration should be given to installation of systems to warn of ventilation failure and oxygen concentrations above 23%.

### Maternity

4.24 For LDRP (labour, delivery, recovery, post-partum) rooms, the diversified flow is based on 10 L/min for the first terminal unit and 6 L/min for 25% of the remainder. Two cot spaces may be provided, each with a terminal unit. Only one will be considered to be in use. The diversified flow for cot spaces is based on 10 L/min for the first and 50% of the remainder at 3 L/min.

4.25 In the event of multiple births, the additional gas usage will have negligible overall effect on the total flow.

4.26 Maternity department operating rooms are designed as a suite; that is, it is presumed that oxygen will be provided either in the anaesthetic room or in the operating room. In post-anaesthesia recovery, it is assumed that 75% of beds will require oxygen to be delivered.

## Hyperbaric oxygen chambers

4.27 Hyperbaric oxygen chambers should be supplied from a separate branch from the main riser/distribution pipe: the pipeline system should be from a liquid supply source. Typical flows for a single patient chamber are as shown in Table 14.

### Table 14 Gas flow – hyperbaric chambers

| | Max. time for one complete treatment | Total consumption for max. treatment time (L) | Consumption for each additional minute (L/min) |
|---|---|---|---|
| $O_2$ atmosphere and recirculation: | | | |
| On open circuit | 2 hours | 30,000 | 250 |
| On recirculation | 2 hours | 7,250 | 40 |
| $O_2$ only, no recirculation | 2 hours | 30,000 | 250 |
| $O_2$ delivery by built-in breathing mask and overboard pump | 2 hours | 1,200 | 10 |
| $O_2$ delivery by built-in breathing hood and overboard pump | 2 hours | 7,250 | 60 |

## Notes

a. The flows for a recirculating unit assume the standard method of operation is recirculation throughout the treatment. It is recommended that the pipeline should be designed for open circuit operation to ensure adequate flow under all conditions.

b. Clinical practice may require the inclusion of air during the treatment; it may also be necessary to switch to air in the unlikely event of an oxygen convulsion. Therefore consideration should be given to the provision of medical air from a separate dedicated medical air plant in accordance with Chapter 7.

c. Some hyperbaric chambers use air as a buffer and consequently less oxygen is consumed. The advice of the manufacturer should be sought. Where this is the case, the air should be supplied from a separate supply system complying with the requirements for medical air systems.

## Nitrous oxide

4.28 Nitrous oxide is provided for anaesthetic purposes and occasionally for analgesic purposes. In all cases, each terminal unit should be capable of passing 15 L/min, but in practice the flow is unlikely to exceed 6 L/min.

4.29 When calculating diversities in a department, 15 L/min is allowed for the first and 6 L/min for the remainder, subject to the appropriate diversity factor being applied (see Table 15).

4.30 It is assumed that, for an operating department, nitrous oxide may be in use simultaneously in all operating rooms. As it is unlikely that a patient would be anaesthetised in the anaesthetic room at the same time that a patient in the associated operating room was continuing to be treated under an anaesthetic (and because the duration of induction is short), no additional flow is included.

## Nitrous oxide/oxygen mixture

4.31 All terminal units should be capable of passing 275 L/min for a very short period (normally of five seconds' duration) to supply inhalationary "gasps" by the patient, and a continuous flow of 20 L/min. The actual flow would not normally exceed 20 L/min.

4.32 The diversified flow in delivery rooms is based on 275 L/min for the first bed space and 6 L/min for each of the remainder, of which only half of the women in labour will be using gas for 50% of the time. (The peak inhalationary "gasp" is 275 L/min, whereas the respirable minute volume will be catered for with a flow of 6 L/min – it should also be borne in mind that a woman in labour would not continuously breathe the analgesic mixture.) For larger maternity departments with twelve or more LDRP rooms, two peak inhalationary "gasps" are included.

4.33 Nitrous oxide/oxygen mixture may be used in other areas for analgesic purposes. The diversified flow is based on 10 L/min for the first treatment space, and 6 L/min for a quarter of the remainder for 25% of the time.

4.34 Design and diversified flows for nitrous oxide/oxygen mixtures are given in Table 16.

**Table 15 Nitrous oxide: design and diversified flows**

| Department | Design flow for each terminal unit (L/min) | Diversified flow $Q$ (L/min) |
|---|---|---|
| Accident & emergency: resuscitation room, per trolley space | 10 | $Q = 10 + [(n - 1)6/4]$ |
| Operating | 15 | $Q = 15 + (nT - 1)6$ |
| Maternity: operating suites | 15 | $Q = 15 + (nS - 1)6$ |
| Radiological: all anaesthetic and procedures rooms | 15 | $Q = 10 + [(n - 1)6/4]$ |
| Critical care areas | 15 | $Q = 10 + [(n - 1)6/4]$ |
| Oral surgery/orthodontic: consulting rooms, type 1 | 10 | $Q = 10 + [(n - 1)6/4]$ |
| Other departments | 10 | No additional flow included |
| Equipment service rooms | 15 | No additional flow included |

**Table 16 Nitrous oxide/oxygen mixtures – design and diversified flows**

| Department | Design flow for each terminal unit (L/min) | Diversified flow $Q$ (L/min) |
|---|---|---|
| **Maternity:**<br><12 LDRP room(s), mother<br>>12 LDRP rooms | 275 | $Q = 275 + [(n - 1)6/2]$<br>$Q = 275 \times 2 + [(n - 1)6/2]$ |
| **Other areas** | 20 | $Q = 20 + [(n - 1)10/4]$ |
| **Equipment service rooms** | 275 | No additional flow included |

## Air

4.35 Air is used to provide power for several types of equipment including surgical tools, ventilators and nebulisers. Oxygen should be avoided as a power source because of fire risk and cost, and should not be used where medical air is available, unless specifically recommended by the device manufacturer.

4.36 Air should be provided at two different pressures but to the same Ph. Eur. standard:

a. a pressure of 400 kPa is required for medical air to drive ventilators and for other respiratory applications;

b. a pressure of 700 kPa or higher is required for surgical air to drive surgical tools.

### Medical air 400 kPa

*General*

4.37 The use of medical air, particularly for respiratory use and during anaesthesia, has increased markedly in recent years. This service is the most critical of the medical gas services, since air-powered ventilators cease to operate in the event of failure of the supply.

4.38 Medical air is also directly inhaled by patients during ventilation. It may also be used to dilute oxygen before administration because of the potentially toxic effects of pure oxygen.

4.39 The supply system for medical air 400 kPa may be a manifold system, a compressor system or a proportioning system (synthetic air), and includes an emergency reserve manifold. A compressor plant, or synthetic air supply, should always be specified where air-powered ventilators are to be used.

4.40 One of the major uses of medical air is for patients' ventilators, which fall into two main categories – those used during anaesthesia and those used during critical care. Pneumatically-powered ventilators can use up to 80 L/min free air continuously. The exact flow requirements will depend on the design of the ventilator. The flow and pressure requirements for some typical ventilators are given in Table 17.

4.41 Current models of anaesthetic ventilator are very similar to critical care models, and may require peak flows of up to 80 L/min and average flows of 20 L/min. Almost all such units are pneumatically driven and electronically controlled.

4.42 Medical air 400 kPa is also used for other equipment such as anaesthetic gas mixers, humidifiers and nebulisers. The flow rates normally required would not exceed 10 L/min, and this flow is always in excess of the actual volume respired.

*Pressure requirements*

4.43 A minimum pressure required at terminal units for respiratory use is 370 kPa.

4.44 Medical air should not be used to supply mechanical services (see paragraph 7.130).

4.45 Some medical gas pendants use the medical air supply for operating the control/retraction system. This is permitted, provided that:

a. a flow limiting device is provided to protect the medical air system in the event of failure of any downstream component;

b. a non-return valve is incorporated to protect the system integrity;

c. appropriate AVSU arrangements are in place (see Chapter 3).

(The surgical air supply should be used to provide the power source whenever possible.)

4.46 The flow requirements should be ascertained and taken into account prior to the installation of the equipment.

### *Flow requirements*

4.47 Flow requirements for medical air are given Table 18. In ward areas and treatment rooms, the use of medical air is most likely to be for nebulisers. In these areas they should be capable of passing 20 L/min, although typically 10 L/min will be required in in-patient accommodation where air is used for nebulisers.

## Table 17 Typical pressure and flow requirements for ventilators and nebulisers

| Ventilator type | Pressure (kPa) | Flow (L/min) |
|---|---|---|
| Anaesthesia, typically gas-driven, electronically controlled | Nominally 400. Max 600(1) | Pneumatically driven ventilators use up to 80 max.<br>20 continuous |
| Critical care, electrically controlled, gas-powered | Nominally 400. Max 600(1) | 180 peak(2)<br>80 continuous |
| Neonatal, gas-driven, electronically controlled | Nominally 400. Max 600(1) | 80 peak(2)<br>40 continuous |
| Nebulisers | 400 | 10 |

Notes:

1. It is strongly recommended that ventilators are not connected to the 700 kPa system since their blenders only work satisfactorily with a tolerance of about 10%: with high differential pressures for air and oxygen an incorrect mixture could be obtained.

2. These flows can be achieved under certain clinical conditions. The peak flows are usually of very short duration.

**Table 18 Medical air 400 kPa – design and diversified flows**

| Department | Design flow for each terminal unit (L/min) | Diversified flow $Q$ (L/min) |
|---|---|---|
| **In-patient accommodation (ward units):** | | |
| Single/multi-bed and treatment rooms[(1)] | 20 | $Q_w = 20 + [(n - 1)10/4]$ |
| Ward block/department | 20 | $Q_d = Q_w[1 + (nW - 1)/2]$ |
| **Accident & emergency:** | | |
| Resuscitation room, per trolley space | 40 | $Q = 40 + [(n - 1)20/4]$ |
| Major treatment/plaster room, per trolley space | 40 | $Q = 40 + [(n - 1)20/4]$ |
| Post-anaesthesia recovery, per trolley space | 40 | $Q = 40 + [(n - 1)40/4]$ |
| **Operating:** | | |
| Anaesthetic rooms | 40 | No additional flow included |
| Operating rooms | 40 | $Q = 40 + [(nT - 1)40/4]$ |
| Post-anaesthesia recovery | 40 | $Q = 40 + [(n - 1)10/4]$ |
| **Maternity:** | | |
| LDRP rooms: | 40 | $Q = 40 + [(n - 1)40/4]$ |
| Baby[(2)] | 40 | $Q = 40 + [(n - 1)40/4]$ |
| Operating suites: | | |
| Anaesthetist | 40 | $Q = 40 + [(nS - 1)10/4]$ |
| Post-anaesthesia recovery | 40 | $Q = 40 + [(n - 1)40/4]$ |
| Neonatal unit (SCBU) | 40 | $Q = 40_n$ |
| **Radiological:** | | |
| All anaesthetic and procedures rooms | 40 | $Q = 40 + [(n - 1)40/4]$ |
| **Critical care areas**[(3)] | 80 | $Q = 80 + [(n - 1)80/2]$ |
| **High-dependency units** | 80 | $Q = 80 + [(n - 1)80/2]$ |
| **Renal** | 20 | $Q = 20 + [(n - 1)10/4]$ |
| **Oral surgery/orthodontic:** | | |
| Major dental/oral surgery rooms | 40 | $Q = 40 + [(n - 1)40/2]$ |
| **All other departments** | 40 | No additional flow allowance to be made |
| **Equipment service rooms** | 40 | No additional flow included |

Notes:

1. It is assumed that a patient will use oxygen in a ward or in the treatment room.

2. Where two cot spaces have been provided in an LDRP room, assume only one will require medical air.

3. This diversified flow is also used for helium/oxygen mixture (see paragraph 4.73).

## Surgical air 700 kPa

4.48 The pressure requirements of surgical tools are between 600 and 700 kPa and flows may vary between 200 and 350 L/min (STP; see Table 19). Most surgical tools are designed to operate within this pressure range. Higher pressures are likely to cause damage to tools. Inadequate tool performance, however, is likely to result from the lack of flow at the specified pressure.

4.49 The introduction of synthetic air (from on-site blending of oxygen and nitrogen) leads to the possibility of using nitrogen as the power source for surgical tools.

4.50 The pipeline systems should be designed to provide a flow of 350 L/min at 700 kPa at the outlet from the terminal unit. Existing systems may not meet this requirement (but should be capable of delivering 250 L/min at the terminal unit).

### Note

Some surgical tools require up to 500 L/min at up to 1400 kPa. These will require a separate supply, normally from cylinders.

### *Dual pressure surgical air systems*

4.51 There are cases where, because of system size, a simple single regulation system (that is, directly from a receiver pressure of say, 10 bar, to a line pressure of approximately 8 bar) will not ensure correct flow conditions at the surgical air terminal units. To overcome possible flow problems, a double pressure regulating system can be used.

4.52 Such a system will involve a compressed air plant receiver operating at a typical pressure of 13 bar, followed by first-stage pressure regulation to a line pressure of 11 bar. Locally-sited pressure regulators (for example for each operating room) are provided to give the recommended flow and pressure at the terminal unit outlet(s).

4.53 If this type of system is installed, for design purposes the maximum allowable pressure drop of 5% should be taken from the plantroom wall to the upstream side of the secondary regulator. The secondary regulator should be adjusted to give 700 kPa at a flow of 350 L/min at the terminal unit outlets(s) and should not allow the static pressure on the upstream side of the terminal unit to rise above 9 bar.

### *Diversity*

4.54 Surgical air 700 kPa is only required where surgical tools are to be used. This would typically be orthopaedic and neurosurgery operating rooms, and possibly plaster rooms. For flexibility, and to allow for possible overspill, surgical air should be extended to two to four adjacent operating rooms. It is not required in maternity or ophthalmology operating rooms.

**Note**

Health Technical Memorandum 2022, Supplement 1 – 'Dental compresssed air and vacuum systems' allows for the extension of surgical air into dental departments for tool use only. No diversity factor should be applied to the dental service as all dental clinics can be in use simultaneously: the total design flow of the dental department should be added.

4.55 Unlike dental departments, the use of surgical tools in an operating procedure takes place for a limited period of time.

**Table 19 Typical pressure and flow requirements for surgical tools**

| Type of tool | Pressure (kPa) | Flow (L/min) |
|---|---|---|
| Small air drill | 600–700 | 200 |
| Medullary reaming machine | 600–700 | 350 |
| Oscillating bone saw | 600–700 | 300 |
| Universal drill | 600–700 | 300 |
| Craniotome | 620–750 | 300 |

### *System capacity*

4.56 Unlike respirable equipment, surgical tools are used intermittently, typically for a few seconds, up to a maximum of three minutes. The plant, therefore, should have the capacity to provide the design flow of the pipeline for a maximum period of five minutes in any 15-minute period. The diversified flow is based on the assumption of 350 L/min for the first theatre and a quarter of the remainder – see Table 20.

### *Terminal units intended for equipment testing*

4.57 It may be necessary to provide surgical air at 700 kPa in the equipment service workshop for testing purposes. Unless a surgical air 700 kPa pipeline is available nearby, it may be cost-effective

**Table 20 Surgical air 700 kPa – design and diversified flows**

| Department | Design flow for each terminal unit (L/min) | Diversified flow $Q$ (L/min) |
|---|---|---|
| Operating room (orthopaedic and neurosurgical operating rooms only):<br><4 operating rooms<br>>4 operating rooms | <br>350<br>350 | <br>$Q = 350 + [(n - 1)350/2]$<br>$Q = 350 + [(n - 1)350/4]$ |
| Other departments, eg equipment workshops, fracture clinic | 350 | $Q = 350$ |
| Equipment service rooms | 350 | No additional flow required |

to use portable cylinders, with a two-stage regulator.

4.58 If a pipeline supply is to be provided, each terminal unit should be capable of delivering 350 L/min. Where several terminal units are provided, it is unlikely that more than one terminal unit will be in use at any time, and therefore the total design flow for the equipment service workshop will be 350 L/min.

## Vacuum

### General

4.59 In virtually all cases, vacuum is used via a suction control device and fluid is collected in suction jars. On wards these are typically of approximately 1 L capacity. In operating rooms, two or four 2–3 L capacity vessels are provided for the suction control regulator.

4.60 Once full, suction jars have to be emptied; therefore, vacuum cannot be applied continuously.

4.61 The greatest generation of fluid to be aspirated is likely to arise in the operating room, particularly during orthopaedic surgery, where jet lavage to irrigate and cleanse the wound may be in use. The maximum rate of collection is about 4 L/min, but it is not continuous.

4.62 During induction of anaesthesia, a patient may vomit. Therefore, it is essential that oral and nasal passages can be cleared as quickly as possible. The highest likely amount of fluid to be aspirated in this case will be no more than 0.5 L.

4.63 In order to aspirate fluid, a suction cannula is normally used, and this will aspirate air as well as the fluid to be removed. The flow required, however, is higher than would be the case if fluid only were to be removed. The ratio of air/fluid aspirated will depend upon the diameter of the cannula.

### In-patient accommodation

4.64 For a 28-bed ward unit comprising single rooms, four-bed rooms and a treatment room, the diversified flow is based on 40 L/min.

4.65 When selecting the size of a sub-branch serving, for example, a four-bed ward, the flow would be taken to be 160 L/min as all four terminal units could, in theory, be in use.

### Medical supply units/bedhead trunking systems

4.66 When designing vacuum (and medical gas systems), it is expected that the greatest pipeline pressure losses will occur near to the terminal units.

4.67 Care must be taken when sizing vacuum pipework within medical supply units with two or more bed/treatment spaces, where availability of space will often limit the size of pipe. The largest size of pipe that can be accommodated (typically 22 mm) should be used, as this will ensure that excessive pressure losses do not occur within the units. Such losses could necessitate the installation of larger diameter pipework within the rest of the system in order to ensure that the system pressure drops prescribed in this Health Technical Memorandum are not exceeded.

4.68 In some instances it may be necessary to provide more than one "feed" from the ceiling distributor to the medical supply unit in order to keep pressure losses within acceptable limits.

### Operating departments

4.69 Vacuum is provided for the surgical team and anaesthetist in the operating room. It is also provided in the anaesthetic and recovery rooms.

4.70 Since it is possible for both the surgical team and anaesthetist to use vacuum simultaneously, each operating room will require 80 L/min and each terminal unit should be capable of passing 40 L/min (see Table 21).

4.71 As it is unlikely that a patient would be anaesthetised at the same time that a patient in the associated operating room was continuing to be treated under an anaesthetic, the need to clear an airway is extremely unlikely and no additional flow is included.

## Helium/oxygen mixture

4.72 Helium/oxygen mixture is used by patients with respiratory or airway obstruction and to relieve symptoms and signs associated with respiratory distress. It can be administered by means of face mask, a demand valve with face mask, a nebuliser, or a ventilator.

4.73 Pipeline supply will be primarily limited to critical care areas, where the gas mixture is used for driving a ventilator. The design and diversified flows should be based on the figures given for medical air (see Table 17).

## Table 21 Vacuum – design and diversified flows

| Department | Design flow for each terminal unit (L/min) | Diversified flow $Q$ (L/min) |
|---|---|---|
| **In-patient accommodation:** | | |
| Ward unit | 40 | $Q = 40$ |
| Multiple ward units | 40 | $Q_d = 40 + [(n - 1)40/4]$ |
| **Accident & emergency:** | | |
| Resuscitation room, per trolley space | 40 | $Q = 40 + [(n - 1)40/4]$ |
| Major treatment/plaster room, per trolley space | 40 | $Q = 40 + [(n - 1)40/4]$ |
| Post-anaesthesia recovery, per trolley space | 40 | $Q = 40 + [(n - 1)40/4]$ |
| Treatment room/cubicle | 40 | $Q = 40 + [(n - 1)40/8]$ |
| **Operating:** | | |
| Anaesthetic rooms | 40 | No additional flow included |
| Operating rooms: | | |
| Anaesthetist | 40 | $Q = 40$ |
| Surgeon | 40 | $Q = 40$ |
| Operating suites | 40 | $Q_s = 80 + [(nS - 1)80/2]$ |
| Post-anaesthesia recovery | 40 | $Q = 40 + [(n - 1)40/4]$ |
| **Maternity:** | | |
| LDRP rooms: | | |
| Mother | 40 | $Q = 40 + [(n - 1)40]/4$ |
| Baby | 40 | No additional flow included |
| Operating suites: | | |
| Anaesthetist | 40 | $Q = 40$ |
| Obstetrician | 40 | $Q = 40$ |
| Operating suites | | $Q_s = 80 + [(nS - 1)80/2]$ |
| Post-anaesthesia recovery | 40 | $Q = 40 + [(n - 1)40/4]$ |
| In-patient accommodation: | | |
| Ward unit comprising single, multi-bed and treatment room | 40 | $Q = 40$ |
| Multi-ward units | 40 | $Q = 40 + [(n - 1)40/2]$ |
| Nursery, per cot space | 40 | No additional to be included |
| SCBU | 40 | $Q = 40 + [(n - 1)40/4]$ |
| **Radiology/diagnostic departments:** | | |
| All anaesthetic and procedures rooms | 40 | $Q = 40 + [(n - 1)40/8]$ |
| **Critical care areas** | 40 | $Q = 40 + [(n - 1)40/4]$ |
| **High-dependency units** | 40 | $Q = 40 + [(n - 1)40/4]$ |
| **Renal** | 40 | $Q_d = 40 + [(n - 1)40/4]$ |
| **Adult mental illness accommodation:** | | |
| ECT room | 40 | $Q = 40 + [(n - 1)40/4]$ |
| Post-anaesthesia, per bed space | 40 | $Q = 40 + [(n - 1)40/4]$ |
| **Adult acute day care accommodation:** | | |
| Treatment rooms | 40 | $Q = 40 + [(n - 1)40/4]$ |
| Post-anaesthesia recovery per bed space | 40 | $Q = 40 + [(n - 1)40/8]$ |
| **Day patient accommodation** (as "In-patient accommodation") | | As "In-patient accommodation" |
| **Oral surgery/orthodontic:** | | |
| Consulting rooms, type 1 | 40 | Dental vacuum only |
| Consulting rooms, types 2 & 3 | 40 | Dental vacuum only |
| Recovery room, per bed space | 40 | $Q = 40 + [(n - 1)40/8]$ |
| **Out-patient:** | | |
| Treatment rooms | 40 | $Q = 40 + [(n - 1)40/8]$ |
| **Equipment service rooms, sterile services etc** | 40 | Residual capacity will be adequate without an additional allowance |

4.74 Helium/oxygen mixtures administered by means of a face mask and cannula, a demand valve with face mask and cannula attached, or a nebuliser, are normally supplied using cylinders fitted with an integral valve.

## Anaesthetic gas scavenging systems

4.75 For anaesthetic gas scavenging systems, it should be assumed that for each operating suite two terminal units could be in use simultaneously, for example in the anaesthetic room and operating room (receiving systems may be left connected when patients are transferred from the anaesthetic room to the operating room). The diversified flows for other departments are as below:

| Department | Design flow for each terminal unit (L/min) | Diversified flow $Q$ (L/min) |
|---|---|---|
| Accident & emergency resuscitation room (per trolley space) | $V$(1) | $Q = V + [(n - 1)V/4]$ |
| Operating departments | $V$ | $Q = V + (nT - 1)V$ |
| Maternity operating suites | $V$ | $Q = V + (nS - 1)V$ |
| Radiodiagnostic (all anaesthetic and procedures room) | $V$ | $Q = V + [(n - 1)V/4]$ |
| Oral surgery/orthodontic consulting rooms (type 1) | $V$ | $Q = V + [(n - 1)V/4]$ |
| Other departments | $V$ | $Q = V + [(n - 1)V/8]$ |

**Note:**

1. For the purpose of sizing the AGS disposal system pump, $V$ is taken as either 130 L/min or 80 L/min (see paragraph 10.16).

# 5 Cylinder manifold installations

5.1 A cylinder manifold installation comprises a primary and secondary supply system.

## Primary supply system

5.2 The primary supply is provided by two banks of equal numbers of gas cylinders which are connected to the pipeline via a control panel. The changeover from the "duty" to the "stand-by" bank of cylinders should be automatic. All manifolds should be capable of passing the full pipeline flow. The temperature of the gas may fall as low as –30°C as the gas passes through the regulator at maximum capacity, and the equipment should be designed accordingly.

5.3 A schematic layout for a primary supply system is shown in Figure 10. Total storage is usually provided on the basis of a risk assessment. Each bank of the manifold should have sufficient cylinders for two days. Additional cylinders for one complete bank change should be held in the manifold room; for nitrous oxide/oxygen mixture, sufficient cylinders to change two banks should be held.

5.4 The nominal and usable capacity of the cylinders commonly used on manifolds are given in Table 22 (the figures are the equivalents at STP).

5.5 An automatic manifold changeover from duty to stand-by should occur at a cylinder pressure that will ensure the greatest possible utilisation of the contents of the cylinders in the duty bank. If the normal operation of the changeover control depends on an electricity supply, the design should be such that failure of the electricity supply does not disrupt the flow of gas to the distribution system.

### Note

Some systems are designed so that both banks (duty and stand-by) supply gas in the event of a power failure.

5.6 In the event of power failure, when the power is restored, the original "running bank" should be on-line, that is, the same bank that was the "running bank" prior to interruption of the supply.

### Note

Some manifolds default to a specific bank following a power failure, regardless of which bank was the running bank prior to interruption of the supply.

NB: Some units may require manual resetting to the original condition.

**Table 22 Capacities of medical gas cylinders used on manifolds**

| Gas | Nominal capacity (L) at 137 bar | Usable capacity (L)[1] |
|---|---|---|
| Oxygen J-size | 6800 | 6,540 |
| Nitrous oxide:<br>J-size<br>G-size | <br>18,000<br>9,000 | <br>–<br>8,900 |
| Nitrous oxide/oxygen mixture G-size | 5,000 | 4,740 |
| Medical air J-size | 6,400 | 6,220<br>5,550 |
| Helium/oxygen mixture K-size | – | 7000 nominal |

Note:

1 The usable figures are for discharge to a gauge pressure of 7 bar. Two sets of figures are provided for air – for 400 kPa systems and 700 kPa systems. The latter is for discharge to 15 bar.

## Figure 10 Typical automatic manifold control system and emergency reserve manifold

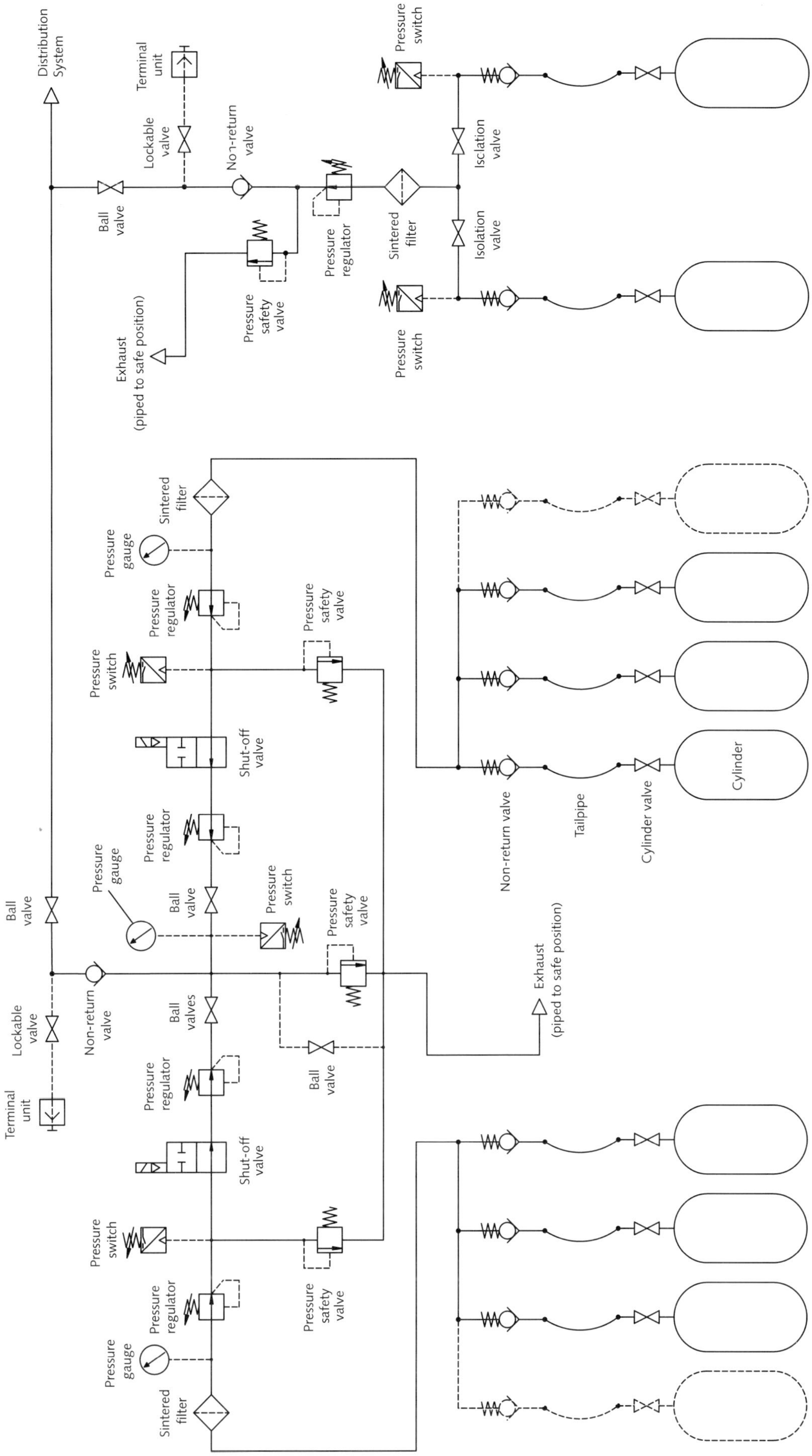


Symbols to BS 2971:1993/ISO 1219-1:1991

5.7 Manifolds and control panels should be designed and certificated for use with 230 bar cylinders. The manifold headers should incorporate a renewable non-return valve to prevent the discharge of a complete bank of cylinders in the event of "tail-pipe" rupture.

5.8 The tail-pipe cylinder connector must be a pin-index yoke connector in accordance with BS EN ISO 407:2004 for oxygen, nitrous oxide/oxygen mixture (50% v/v) and medical air. No non-metallic flexible connectors should be used. The connector for nitrous oxide should be a side outlet valve connector in accordance with BS 341-3:2002. The manifold connectors should be in accordance with the following:

| Thread | Medical gas |
|---|---|
| M24 x 2 | Medical air |
| M22 x 2 | $N_2O/O_2$ |
| M20 x 2 | $O_2$ |
| M18 x 2 | $N_2O$ |

Where it is necessary to use non-metallic materials, consideration should be given to the use of non-halogenated polymers in high pressure systems (>3000 kPa) delivering oxygen or gaseous mixtures with oxygen concentrations greater than that in ambient air. Consideration should also be given to fitting sintered filters upstream of non-metallic materials to minimise the risk of particle collisions and impacts, which are a potential source of ignition. In addition, there are tests that should be conducted to ensure that the risk of ignition is minimised. Attention is drawn to BS EN ISO 15001: 2004.

### Note

Studies have shown that inadvertent ignition of halogenated polymers can lead to highly toxic by-products being delivered to the gas stream.

5.9 Pressure indication should be provided to indicate pressure in each cylinder bank and in the MGPS.

## Pressure control

5.10 The pressure control should maintain the nominal pipeline pressure within the limits given in Chapter 4. High-pressure regulators should comply with BS EN 738-2:1999 and be supplied with auto-ignition test results.

5.11 Separate pressure regulating valves should be provided for each cylinder bank. The control system should be designed so that the cylinders of one bank can be changed, or the pressure regulator for one bank can be overhauled, without loss of continuity of the gas supply.

5.12 Pressure safety valves should be of the self-closing type and be installed on each distribution pipeline downstream of the manifold line pressure regulator and upstream of the main isolation valve. A pressure safety valve should also be installed between the secondary source of supply (emergency/reserve manifold) and the pipeline distribution system. It should have a flow capacity at least equal to that of the pressure regulator immediately upstream of it. The discharge pipe should be at least one size larger than the main pipeline and be separate for each safety valve.

5.13 This discharge pipeline should be vented to atmosphere, outside the building, in an area where the discharge of oxygen, nitrous oxide, or nitrous oxide/oxygen mixture will not present a fire hazard or cause injury to personnel. Medical and surgical air may be vented internally provided that this is done in a safe way. Warning signs should be posted at the discharge positions; access for inspection should be provided.

5.14 It should terminate at least 3 m clear of any door/window that can be opened or other ventilation/air intake. The ends of the discharge pipelines should be turned downwards to prevent the ingress of dirt and moisture, and be placed and protected so that frost cannot form or be collected upon them. Similar safety valve arrangements are required for installations supplied from liquid oxygen cylinders.

### Note

High pressure cylinders with integral pressure regulation can be used on manifold systems.

## Manifold monitoring and indicating system

5.15 The monitoring and indicating system should perform the following functions:

a. overall manifold monitoring;

b. manifold condition indication;

c. overall supply plant indication.

5.16 All functions should be appropriately identified. Indicators should have a design life of at least five years. The system should be capable of automatic reinstatement after restoration of the power supply.

5.17 Manifold monitoring, indicating and alarm systems should be on the essential electrical supply.

### Manifold control unit

5.18 The control unit should include a green "mains supply on" indicator.

### Manifold monitoring

5.19 Each automatic manifold should be provided with monitoring to detect:

a. duty bank operating;

b. duty bank empty and stand-by bank operating;

c. stand-by bank below 10% capacity, when the duty bank is empty;

d. each secondary supply (emergency reserve) manifold bank below nominal 14 bar (for nitrous oxide) and below 68/100 bar pressure for other gases;

e. pipeline pressure faults outside the normal range.

### Manifold indicator unit

5.20 There should be indicators to show the following conditions:

a. for each automatic manifold:

(i) a green "running" indicator for each bank. This should display when the bank is supplying gas, irrespective of the pressure;

(ii) a yellow "empty" indicator for each bank when the running bank is empty and changeover has occurred;

(iii) a yellow "low pressure" indicator for each bank to be illuminated after changeover, when the pressure in the bank now running falls to the low pressure setting;

b. for each secondary supply (emergency reserve) bank, a yellow indicator to be illuminated when the pressure in the bank falls below 14 bar for nitrous oxide or below 68 bar for other gases (this will require the use of separate pressure sensors – one for each bank);

c. for the pipeline distribution system, a red "low pressure" and a red "high pressure" indicator to be illuminated when the respective conditions occur.

### Alarm signal status unit

5.21 The following indication of manifold conditions should be provided:

a. green "normal": *normal*;

b. yellow "duty bank empty, stand-by running": *change cylinders*;

c. yellow "duty bank empty, stand-by low": *change cylinders immediately*;

d. yellow "emergency reserve bank low": *reserve low*;

e. red "pipeline pressure fault": *pressure fault*.

5.22 Conditions (b) to (e) should be transmitted to the central alarm system. Where relays are used, they should be normally energised relays, which de-energise under fault conditions, with contacts having a minimum rating of 50 V dc and 50 mA. Volt-free, normally closed contacts rated at 50 V dc and 50 mA should be provided for transmission of conditions (b) to (e) to the alarm system.

5.23 The panel can be incorporated into the manifold control unit or be a separate unit within the plantroom. If mounted separately, the cabling should be monitored for open/short circuit. In the event of such a cable fault, a red "system fault" lamp should be illuminated on the alarm signal status unit, together with the appropriate alarm condition.

### Manifold management

5.24 Connections should be provided that allow monitoring of manifold alarm conditions (b) to (e) and manifold running for each "bank". These connections should be volt-free contacts normally closed for each condition having a minimum rating of 50 V dc and 50 mA. The building management system should not be used to control the manifold.

## Secondary supply system

5.25 An emergency reserve manifold system should be provided to form a secondary source of supply, for emergency use, or to permit servicing or repair.

5.26 The supply should be designed to provide the design flow of the primary system and have sufficient connected capacity to supply the pipeline for at least four hours. When such provision would result in more than ten cylinders on each bank, the additional cylinders should be held in the manifold rooms. A non-return valve and isolating valve should be installed immediately upstream of the reserve manifold connection to the pipeline distribution system.

5.27 The requirements for the emergency reserve supply capacity should be set out in the operational policy document; this should take into account the arrangements for the supply of cylinders and the flow that the system is required to provide. The gas supplier should be consulted.

5.28 The specific requirements will depend on the method of primary supply. Where this results in an unrealistic number of cylinders being kept on site, the operational policy should be set out, giving details of procedures to be followed in an emergency to ensure continuity of supply.

5.29 For large installations, it may be impractical to rely on a cylinder manifold system; thus, consideration should be given to either a bulk liquid or liquid cylinder emergency/reserve supply.

5.30 The operational policy document should set out the location of emergency manifolds, cylinders etc and the action to be taken in the event of loss of the primary source of supply.

## Location of secondary sources of supply for manifold installations

5.31 The secondary supply system for cylinder manifold systems should normally be located in the manifold room of the primary supply. A two-cylinder emergency reserve supply would normally be considered adequate for a cylinder manifold supply system (but see note (h) to Tables 1–9 in Chapter 2). All cylinder valves should be permanently open so that gas is immediately available, but one of the isolating valves should be closed. A typical system is shown in Figure 11.

5.32 The supply system should go into operation automatically via a non-return valve, and the emergency reserve manifold (ERM) isolating valve should remain open.

**Figure 11 Typical emergency reserve manifold**

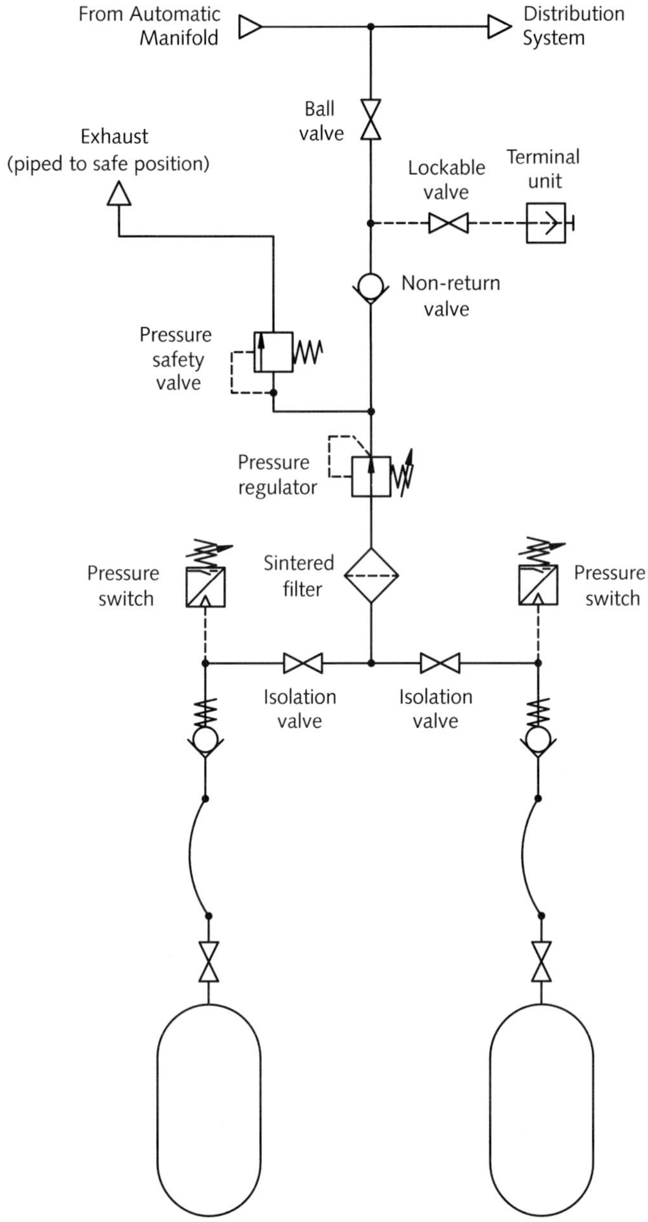


## Secondary sources of supply for air compressors/ liquid oxygen/oxygen concentrators (PSA)

5.33 The supply should comprise a two-bank fully-automatic manifold system as described in paragraphs 5.1–5.23 (except for (d) and (e) in paragraph 5.19; (b) and (c) in paragraph 5.20; and (d) and (e) in paragraph 5.21, which do not apply). A typical number of cylinders in each bank would be a minimum of ten depending on size and location (see also Chapter 2). The manifold system(s) should be installed in an appropriate manifold room(s) separate from the plant.

# 6 Oxygen systems

## Liquid oxygen systems

### General

6.1 Over the last ten years, there has been a significant increase in the use of medical oxygen for treating patients in healthcare facilities, with some hospitals seeing annual increases well in excess of 10%. Introduction of the new European Standard on medical gas pipeline systems (BS EN 737) also has implications.

6.2 The scope of the advice provided by this chapter covers the supply of liquid oxygen to healthcare facilities from delivery into bulk storage vessels for the larger hospitals to supply of liquid oxygen in liquid cylinders to hospitals with lower demands. The guidance given is intended to cover all the aspects of a bulk cryogenic liquid system (VIE).

6.3 This chapter also covers the supply of medical oxygen from on-site generation using PSA plant, or in medical cylinders, other than (in the case of the latter) as a means of backup to the main supply system.

6.4 It does not specifically cover bulk liquid nitrogen installations, but its principles may be applied to hospitals where these gases are used in sufficient quantities to make the use of a VIE cost-effective.

6.5 Significant changes since publication of Health Technical Memorandum 2022 (1997) include:

- the use of risk assessment as a tool to assist in the development of the medical oxygen installation;
- adoption of the principles outlined in BS EN 737-3:2000;
- new methods of sizing the medical oxygen vessels and back-up manifolds;
- designation of vessel contents as "operational" or "reserve" stock.

A checklist for planning and upgrading an oxygen system is given in Appendix H.

*Risk assessment*

6.6 Risk assessment is used to assist in the development of the medical oxygen installation to produce a safe and practical design and ensure that a safe supply of oxygen is available for patient use at all times. It is used for all aspects of the process; from the initial concept designs through installation and operation to the routine assessment of the installation, once in service.

6.7 Advice is given on setting up risk assessment teams and choosing the correct mix of personnel to ensure that all aspects of the associated risks are considered.

6.8 Throughout this chapter, non-exclusive risk criteria lists are provided to assist these teams in identifying the unacceptable risks and suggesting how they might be addressed. It recommends that annual risk assessments are carried out throughout the life of the system to ensure that a safe system of supply is maintained and any new risks are identified.

6.9 The prime responsibility to ensure that adequate stocks of medical oxygen are available for patient use should remain firmly with the hospital's management team. However, the hospital may agree with its gas supplier or facilities management supplier that they should manage the supplies of medical oxygen and maintain adequate stocks in the vessel. These arrangements should be clearly documented within the MGPS operational policy and procedures document. The effectiveness of these arrangements will need to be assessed as part of the risk assessment review and be validated to ensure that they can be met.

6.10 Consideration should be given to the operational management consequences of using different suppliers to supply medical oxygen to different supply systems on the same pipeline system.

6.11 Any contracts involving different suppliers should clearly state the obligations and limitations of liabilities; and any facilities management agreement between the hospital and the medical gas supplier must define the responsibilities of each party.

6.12 There must be no modification to the design or any part of the medical liquid oxygen system without written authorisation from the gas supplier.

### *BS EN 737-3:2000*

6.13 This guidance adopts the principles outlined in BS EN 737-3:2000 'Medical gas pipeline systems: pipelines for compressed medical gases and vacuum', which introduces to the UK the concept of having three independent sources of supply for medical gas systems. This is covered in Chapter 2.

### *New methods of sizing*

6.14 New methods of sizing the medical oxygen vessels and back-up manifolds are covered, together with advice on appropriate location of vessels on site using principles of risk assessment. This ensures the provision of a secure source of supply that reflects the degree of risk associated with the hospital's location and its level of dependency on medical oxygen.

### *Designation of vessel contents as "operational" or "reserve" stock*

6.15 The operational stock is the volume of product that the gas supplier uses to manage deliveries to the hospital; when this stock is exhausted, the vessel should be refilled under normal conditions.

6.16 The reserve stock is the volume of product that is used to provide additional stock to take account of fluctuations in demand or when the supplier fails to make a scheduled delivery.

6.17 Both operational and reserve stock levels are calculated using the risk assessment principles embodied within this document (see Figures 12–17).

## Figure 12 Primary supply (VIE)

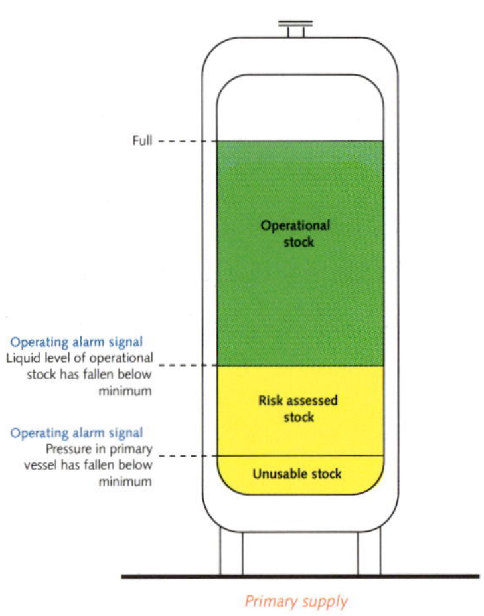


## Figure 13 Primary supply (liquid cylinder)

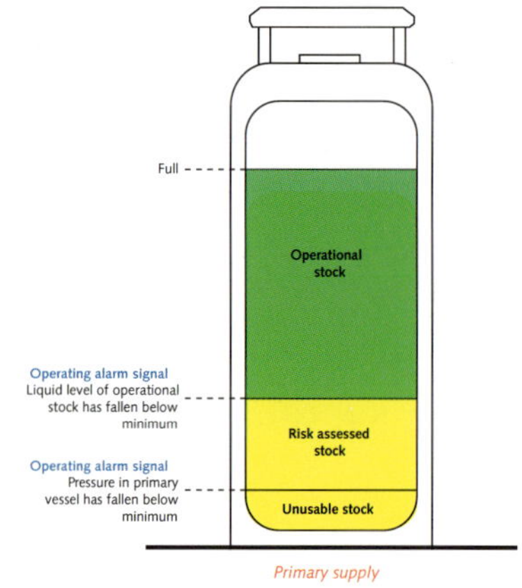


## Figure 14 Secondary supply (VIE)

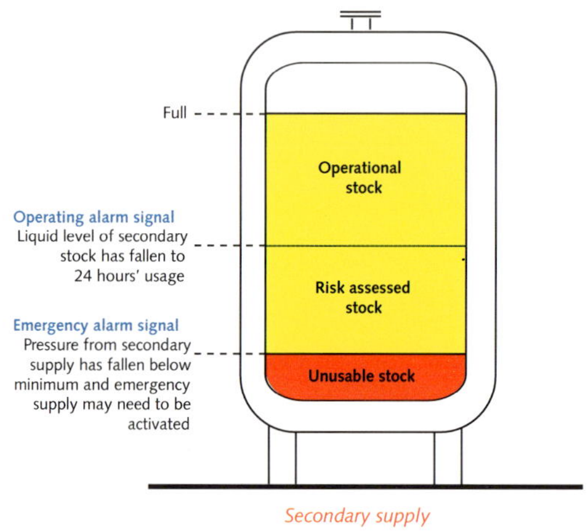


## Figure 15 Secondary supply (liquid cylinder)

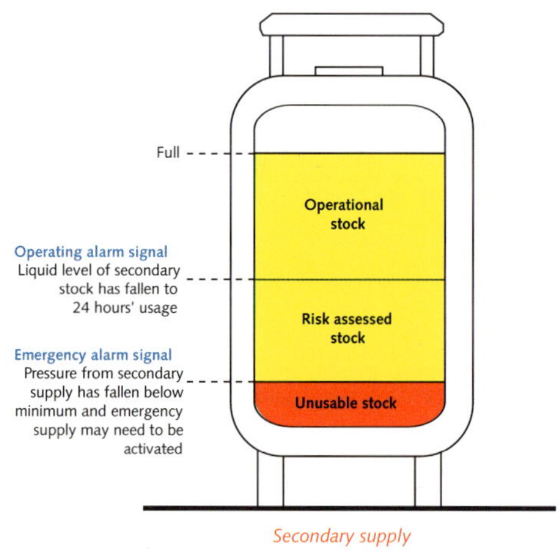

## Figure 16 Secondary supply cylinder manifold

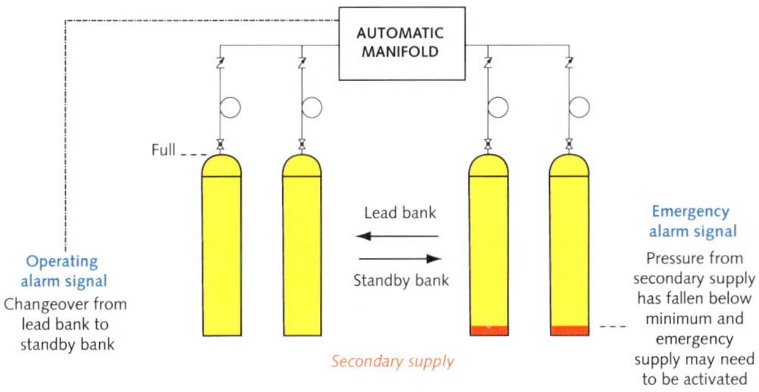


## Figure 17 Secondary supply liquid manifold

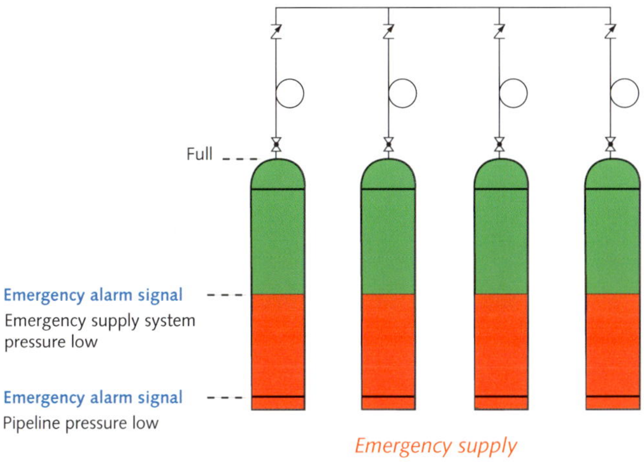


## Choosing an oxygen supply system

6.18 When designing or reviewing an installation to supply medical oxygen to a hospital, the most appropriate method of supplying the gas will be determined by the potential size and variability of the hospital's medical oxygen demand.

6.19 To determine the most suitable and cost-effective method of supplying medical oxygen and the appropriate size of the installation, comprehensive demand figures should be provided to the designer.

6.20 These demand figures (prepared as a part of the risk assessment) should be based on:

- the current average daily gas usage based on the past twelve months' supplies;
- the maximum potential daily demand volumes based on peak flow conditions, as below;
- any planned extensions to the hospital/pipeline that may affect the demand;
- the expected natural annual growth in the use of medical oxygen.

6.21 The maximum potential daily demand should be based on the peak flow conditions measured between 8.00 am and 6.00 pm, with all operating rooms in use and with maximum demand being provided to pipeline outlets. It should not be based on the theoretical pipeline design flow conditions. Where actual flow monitoring is impracticable, daily cylinder or liquid consumption figures should be used.

### Note

Control panels should be capable of passing the maximum design flow of the oxygen system's pipeline distribution system. This may necessitate installing two control panels in parallel.

6.22 Additionally, historic consumption records should be reviewed to assess the current usage and the natural growth of the medical oxygen demand. The growth predictions should take into account any planned extensions to the hospital's facilities or pipeline systems and changes in clinical practices in the hospital that could affect the medical oxygen demand. Natural growth in usage of medical oxygen, due to changes in clinical practice, is about 8% to 10% per annum, but individual hospitals will need to establish this growth figure during the risk assessment process.

6.23 For new hospitals, where no historic information is available, the estimated demand should be based on the proposed size and type of the hospital and the usage figures of the facilities being replaced.

6.24 It is essential to periodically review the average daily demand with the gas supplier and agree either to revise delivery frequencies to maintain the operational stock levels or increase the size of the storage system on site. Any planned increase in demand due to hospital site developments, pipeline extensions or changes in clinical practice should be notified to the gas supplier to ensure that the changes do not jeopardise security of supply.

6.25 The medical liquid oxygen demand should be reviewed with the gas supplier at least annually (or after a significant extension to the pipeline causing increase in demand) to re-assess the size of the installation.

6.26 As the agreed stocks used for the supply of liquid oxygen are all based on an average daily demand, as the demand grows so the storage volume requirements will increase. With the increased volume requirements for the reserve stock, the volume available for operational stock will reduce. Having reviewed the average daily demand with the gas supplier, it is necessary to agree either revised delivery frequencies to maintain the operational stock levels or to increase the size of the storage system on site.

6.27 The review of the medical liquid oxygen installation should also include a review of the risk assessment to ensure that no other conditions on site have been changed that jeopardise the security of the gas supply.

6.28 For smaller hospitals, where the demand is typically below 3000 $m^3$ per annum, the most cost-effective method of supplying medical oxygen is from a compressed gas cylinder manifold.

6.29 As the demand increases, it becomes less practicable to use compressed gas cylinders and more cost-effective to use medical liquid oxygen. A cylinder manifold larger than 2 x 10 J cylinders is likely to prove impracticable because of the manual handling difficulties with the number of cylinders involved. Liquid cylinders, which are ideally suited to an annual consumption of between 3000 $m^3$ and 40,000 $m^3$, can be connected together by a manifold to provide adequate storage capacity and flow rate.

6.30 For hospitals with larger demands, a bulk medical oxygen VIE will generally be used. There is a nominal overlap of annual consumption between 27,500 $m^3$ and 40,000 $m^3$, where either a bulk VIE or a liquid cylinder installation could be considered, either to satisfy a particular requirement, or to accommodate possible site restrictions.

6.31 The main benefit of using gas cylinders is that installation costs of manifold systems are significantly lower than those of a liquid oxygen system. However, the cost of the medical oxygen in compressed gas cylinders is higher than the cost of medical liquid oxygen (supplied either into liquid cylinders or into a VIE). As the demand grows, so the lower unit cost of the liquid oxygen offsets the higher installation costs of the liquid oxygen systems.

6.32 Cryogenic liquid systems are normally used where the demand is high enough to make bulk supplies cost-effective and where the demand makes cylinder supplies impracticable.

6.33 Liquid oxygen provides a flexible approach to both the size and the choice of installation design. Its provision is determined by factors such as the size of the hospital, the availability of space for both the installation and the delivery vehicle, the proximity of the gas supplier and the size of the demand for medical oxygen.

6.34 There are a number of operational benefits in using a medical liquid oxygen system over compressed gas cylinders, including:

- greater volume of medical oxygen stored on site;
- improved security of supply;
- reduced storage area for the medical gas cylinders;
- reduced manual handling requirements for cylinder handling.

6.35 When determining the cost-effectiveness of specific proposals from suppliers, the total supply costs should be assessed, including costs for the site preparation and vessel installation, vessel rental and liquid supply over the total period of the contract.

## System configurations

6.36 In order to comply with the requirements of BS EN 737-3:2000, it is necessary for all medical oxygen installations to have three independent supply sources capable of feeding medical oxygen to the pipeline.

6.37 These three sources are referred to as:

- the primary supply – the main source of medical oxygen on site, providing gas to the pipeline;
- the secondary supply – the secondary source of medical oxygen on site, providing gas to the pipeline and capable of providing the total oxygen flow requirement in the event of a primary supply failure;
- the reserve supply – the final source of supply to specific sections of the pipeline, capable of meeting the required demand in the event of failure of the primary and secondary supplies, or failure of the upstream distribution pipework.

6.38 For smaller hospitals, the primary supply can be fed from compressed gas cylinders but as the demand grows, the most practicable supply source will be either liquid cylinders or a VIE system.

6.39 A fully automatic gas cylinder manifold will normally be used as the secondary supply system for smaller VIEs and liquid cylinder systems. Where it is impracticable to maintain supplies to the hospital using a cylinder manifold, a secondary liquid oxygen system will be necessary.

6.40 Emergency supplies will not normally be fed from a liquid oxygen supply system, as it is not possible to prevent the boil-off of the liquid oxygen over extended periods when the emergency system is not in use.

6.41 All attempts should be made to locate the primary and secondary supply systems on separate sites. They should have independent control systems and supply routes into the hospital pipeline system and be valved accordingly to ensure that the systems remain independent.

6.42 Where it is not feasible to utilise two sites, the risk assessment should evaluate the greater level of risk associated with using a single site and define the appropriate actions that should be established to obviate the higher risks, such as using twin or ring-main pipeline systems, siting of the emergency supply manifold or installing suitable protection for the installation.

6.43 The overall system should be designed so that the primary supply is used first, with the secondary supply automatically switching in when the primary supply is either empty or fails to supply.

## Primary supply systems

### *Cryogenic liquid systems (VIE)*

6.44 These systems, commonly referred to as vacuum-insulated evaporators (VIEs), are used to store the medical gas as a liquid at cryogenic temperatures and to vaporise it into a gas at ambient temperature for distribution through the hospital pipeline.

*Plant*

6.45 The VIE system consists of:

- a vacuum-insulated cryogenic storage vessel to store the bulk liquid;
- one or more ambient-heated vaporisers to convert the cryogenic liquid into a gas for supply to patients via a pipeline;
- control equipment to control the pressure and flow of gas to the pipeline.

6.46 The liquid oxygen is stored at cryogenic temperatures (down to minus 183°C) and converted to a gas at ambient temperature by passing it through an air-heated vaporiser.

6.47 The cryogenic storage vessel is normally constructed from a stainless steel inner pressure vessel that is supported in a mild steel outer shell. The space between the vessels is filled with a high-performance insulating material, maintained under a vacuum, to minimise heat transfer to the inner vessel, which reduces the rate of evaporation of the liquid oxygen.

6.48 A pressure-raising regulator that permits the flow of liquid to the pressure-raising vaporiser, as required, automatically controls the pressure in the liquid oxygen system. The vaporised liquid is fed back to the top of the vessel or liquid cylinder to maintain the pressure in the system.

6.49 Under normal operating conditions for a VIE system, the gas supply to the hospital will be maintained by feeding liquid oxygen to the main vaporiser system where it is converted to a gas and warmed towards ambient temperature. There is a tendency for the vaporiser system to "ice up" where hospital demands are high or continuous, or airflow to the vaporisers is restricted. Under these circumstances the options to be considered should include:

- installation of additional vaporisers;
- an auto-changeover system between vaporisers;

**Note**

In the event of power failure, control valves on all vaporisers should fail "open".

- hot water/electrically heated vaporisers;
- increasing size of vaporiser;
- repositioning.

6.50 Where hospital demands are low or very erratic, the natural heat transfer into the vessel causes the liquid oxygen to boil and the vessel pressure to rise. When the vessel pressure rises to a set point, the hospital pipeline is fed from the top gas to prevent the vessel pressure rising above the safety-valve setting. On safety-valve operation, oxygen must be able to vent safely to atmosphere.

6.51 In all cases, the pipeline pressure is controlled using a system of duplex pressure regulators and valves. It is essential that all materials used in the construction of the vessels, control equipment and pipeline are compatible with oxygen at the operating temperatures that could be encountered under normal operation with single fault condition.

The risk assessment will determine the exact configuration.

6.52 The control panel design should comply with the design requirements specified in BS EN 737-3 and be sized to provide the system design flow.

*Telemetry*

6.53 The use of telemetry on the liquid storage system is recommended because it permits both the hospital and the gas supplier to monitor relevant supply conditions continuously, including storage vessel levels and pressure. In addition, it can be used to transmit other operational data from the storage vessel, pipeline and associated equipment for monitoring purposes. It may be beneficial to make this information available to the relevant person(s) in the trust. By having continuous monitoring of stock available through the telemetry system, an existing vessel could be retained. This solution is only acceptable provided that an appropriate risk assessment, following the guidance given in this chapter, supports the decision.

*Siting*

6.54 The Authorised Person (MGPS) will be responsible for agreeing the final location of the liquid oxygen compound(s), taking into consideration any issues raised in the initial risk assessment.

6.55 When considering the space requirements for the liquid oxygen compound(s), there may be operational advantages in having two compounds in different areas on the hospital site, rather than one larger site utilising either a single large vessel or multiple tanks. This arrangement may also have benefits with respect to both planning permission and meeting the safety distances specified in the British Compressed Gases Association's (BCGA) Code of Practice 19 (CP19): 'Bulk liquid oxygen storage at users' premises' (henceforth known as BCGA CP19; see Table 23).

6.56 Each supply system should be located in a secure fenced compound, which should be sized to allow adequate access to all of the control equipment.

6.57 The site should essentially be level but designed to have adequate falls to prevent water accumulating beneath equipment.

6.58 The location of drains in the vicinity of the site should comply with the requirements specified in BCGA CP19 (see Table 23).

6.59 Only under extreme conditions should the safety distances specified in BCGA CP19 be reduced. Any relaxation of these safety distances needs to be agreed with the supply company's safety representative and the Authorised Person (MGPS). Both parties must ensure that an equivalent level of safety is achieved, and this should be approved and documented.

6.60 The layout of the liquid oxygen installation should provide adequate access to all of the relevant components of the system and permit adequate airflow for the ambient vaporisers.

6.61 The plinth should be of concrete construction. The area in front of the vessel(s) (tanker apron) should be non-porous concrete. Under no circumstances should tarmac be used in the vicinity of the liquid oxygen filling point, or areas where liquid oxygen spillage may occur.

6.62 The location of the liquid oxygen compound should permit the supplier to gain safe access with the appropriately sized tanker. It is the supplier's responsibility to assess the space requirements for vehicular access.

6.63 The design of the liquid oxygen installation should take into account the gas supplier's requirements for discharging the liquid oxygen from the cryogenic tanker. The area directly in front of the vessel should be kept clear to provide access for the delivery vehicle at any time. Under no circumstances should cars be permitted to park in front of the compound.

6.64 The compound should not be used for the storage of other equipment.

6.65 Where the secondary or emergency supply system is fed from a cylinder manifold, it should be in a separate enclosure/manifold room and have adequate space to permit safe cylinder changeover. Spare cylinders should not be held in the VIE compound or liquid cylinder compound but stored in the nearest medical cylinder store.

6.66 A pipework and installation diagram (P&ID) of the plant should be displayed clearly to indicate the appropriate valves that are necessary to operate the plant safely. The medical gases supplier should make the Authorised Person (MGPS), and others in the hospital that may operate the system, aware of its general operating principles. (Typical plant installations are shown in Figures 18–20.)

## Table 23 Safety distances to comply with BCGA CP19

| Safety distances from exposure to vessel/point where oxygen leakage or spillage can occur | Up to 20 tonnes (metres) | Over 20 tonnes (metres) |
|---|---|---|
| Areas where open flames/smoking permitted | 5 | 8 |
| Places of public assembly | 10 | 15 |
| Offices, canteens and areas of occupancy | 5 | 8 |
| Pits, ducts, surface water drains (untrapped) | 5 | 8 |
| Openings to underground systems | 5 | 8 |
| Property boundaries | 5 | 8 |
| Public roads | 5 | 8 |
| Railways | 10 | 15 |
| Vehicle parking areas (other than authorised) | 5 | 8 |
| Large wooden structures | 15 | 15 |
| Small stocks of combustible materials, site huts etc | 5 | 8 |
| Process equipment (not part of installation) | 5 | 8 |
| Continuous sections of flammable gas pipelines | 3 | 3 |
| Flanges in flammable gas pipelines (over 50 mm) | 15 | 15 |
| Fuel gas vent pipes | 5 | 8 |
| Compressor/ventilator air intakes | 5 | 8 |
| Fuel gas cylinders (up to 70 $m^3$) | 5 | 5 |
| LPG storage vessels (up to 4 tonnes) | 7.5 | 7.5 |
| LPG storage vessels (up to 60 tonnes) | 15 | 15 |
| Bulk flammable liquid storage vessels (up to 7.8 $m^3$) | 7.5 | 7.5 |
| Bulk flammable liquid storage vessels (up to 117 $m^3$) | 15 | 15 |
| MV or HV electrical sub-stations | 5 | 8 |

## Figure 18 Primary supply VIE system with compressed gas cylinder manifold

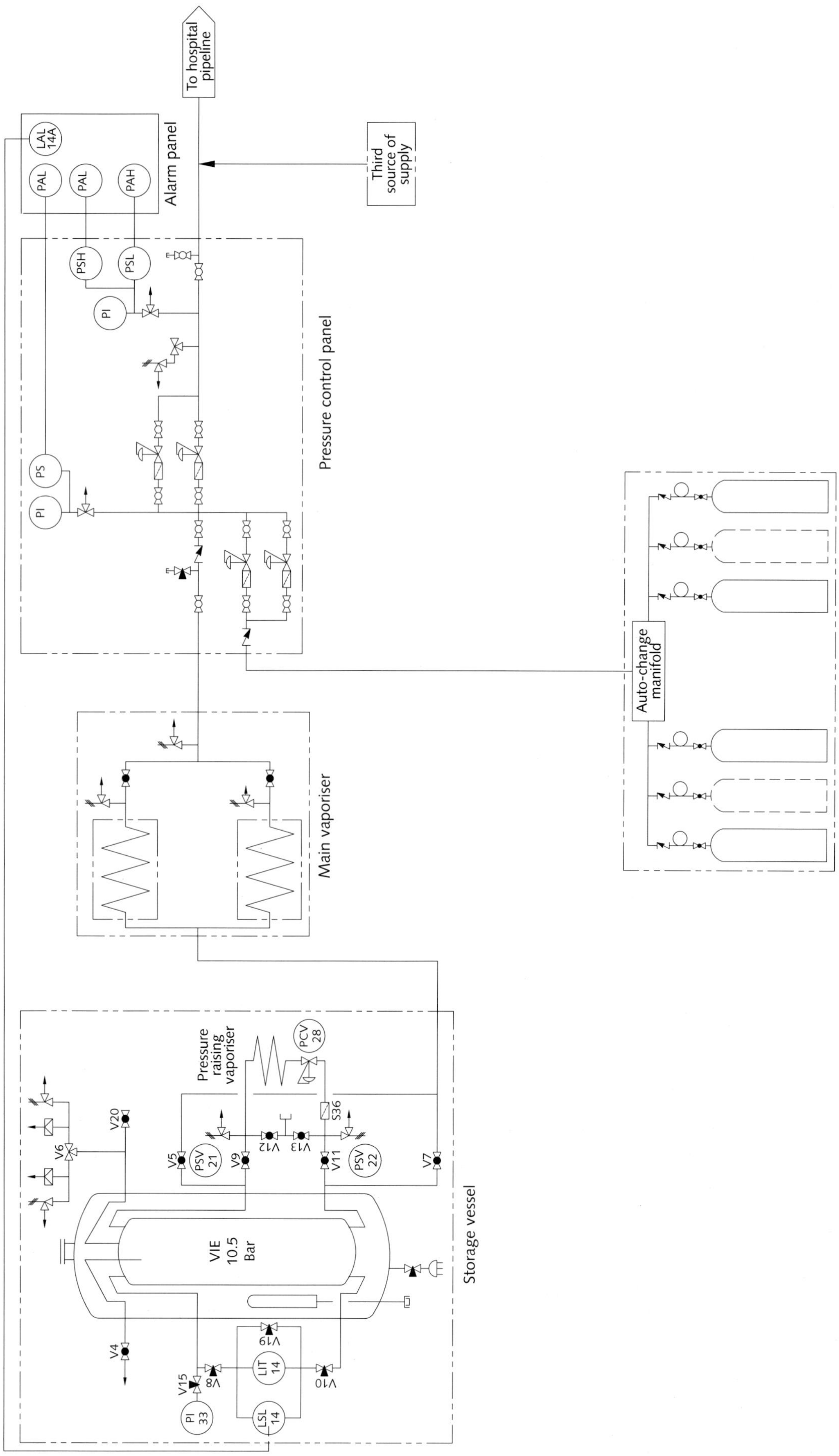

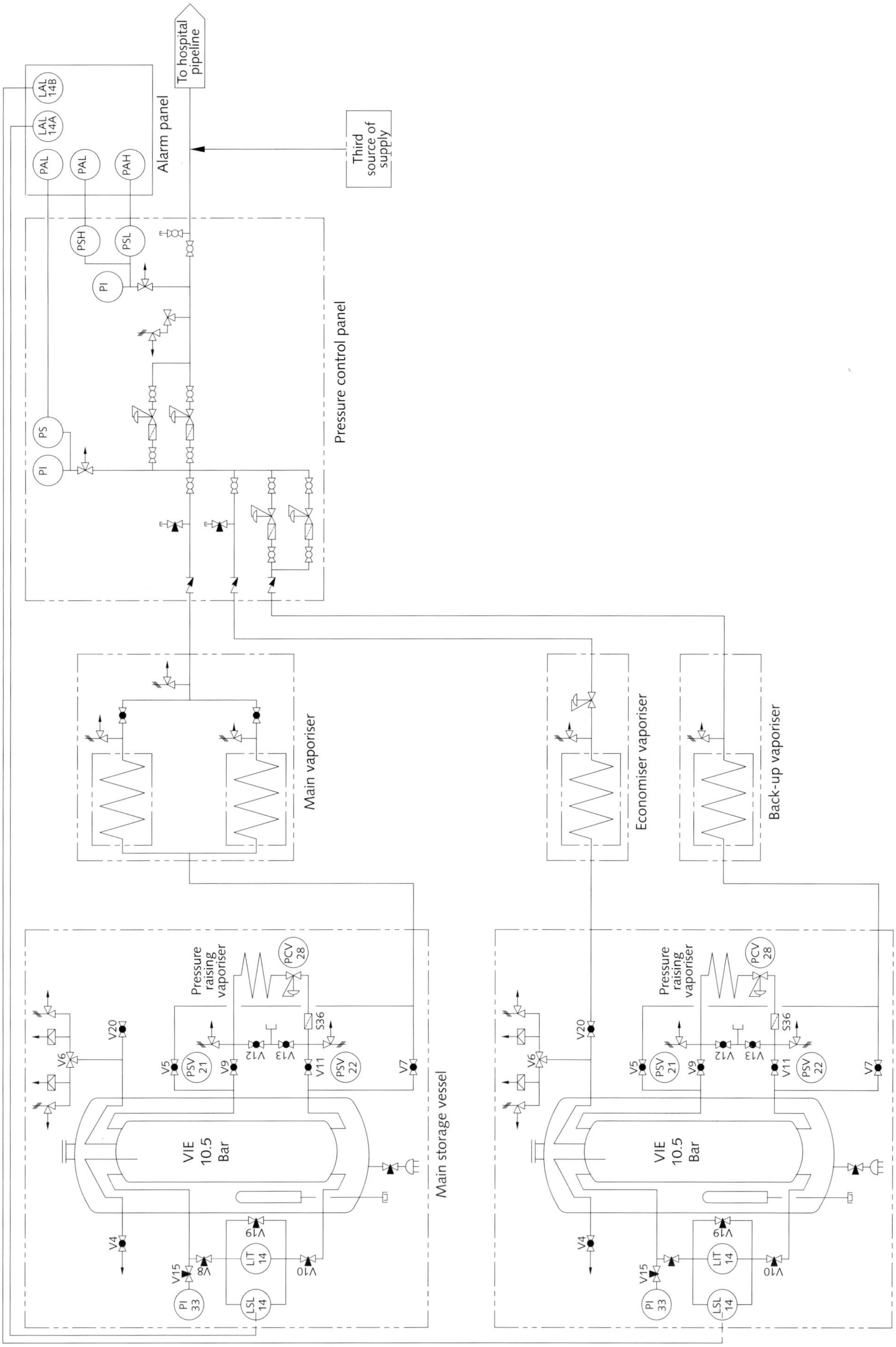


Figure 19 Primary and secondary VIE system on single plinth

## Figure 20 Primary and secondary VIE system on separate plinth

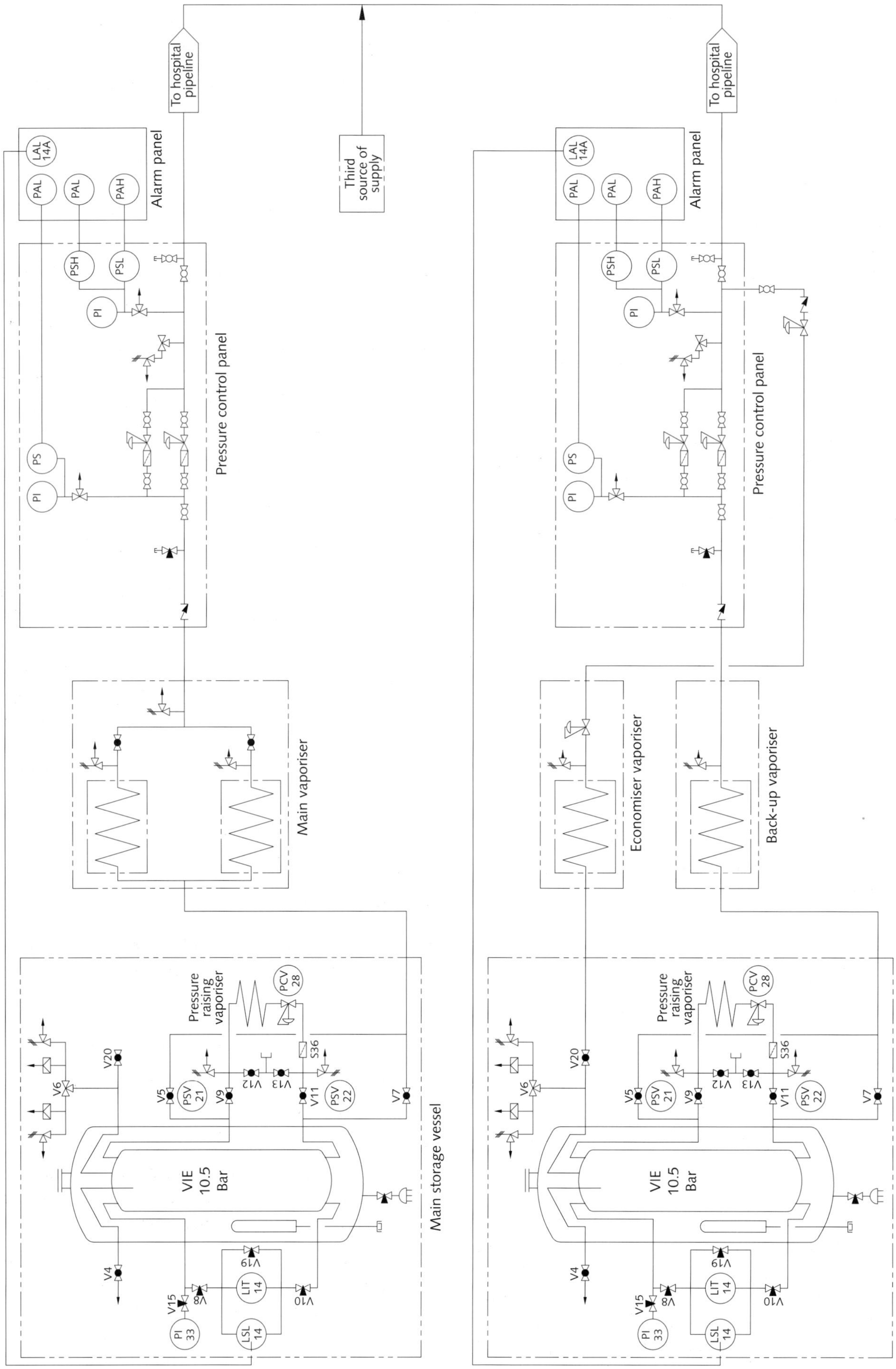

### *Liquid cylinder systems*

*Plant*

6.67 Liquid cylinder systems can also be used to store the medical gas as a liquid at cryogenic temperatures and to vaporise it into a gas for patient use. These systems are used where the demand is too high for compressed gas cylinders to be a practicable option but where it is neither economic nor possible to supply bulk medical liquid oxygen in a VIE system.

6.68 Liquid cylinders are constructed in a similar way to vacuum-insulated cryogenic storage units, that is, as double-walled vessels. However, unlike the VIE, the liquid cylinder has an internal vaporiser coil to convert the liquid into a gas.

6.69 The size of the liquid cylinder can vary between 200 L and 1000 L water capacity. To obtain sufficient storage capacity and to meet the hospital's flow requirements, a number of liquid cylinders can be connected together via a manifold.

6.70 The liquid cylinder system consists of:

- a number of vacuum-insulated liquid cylinders;
- a system to manifold the liquid cylinders together to store sufficient liquid on site to meet the hospital's demand;
- control equipment to regulate the pressure and flow of gas to the pipeline.

6.71 Although liquid cylinders are suitable for transportation when full, they are normally installed as a fixed installation and remotely filled whilst in situ.

6.72 Liquid cylinders are designed and supplied with gas-specific liquid-fill and gas-use connections (including the connection on the remote liquid-fill connection where the liquid cylinders are filled in situ).

6.73 The connections used are:

- liquid fill: CGA 440;
- gas use: ISO 5145.

*Siting*

6.74 Where there is no alternative, a liquid cylinder manifold may be installed in a building or confined area, but only if the vent header (to which all liquid cylinder vents will be connected) is piped to a safe area via a back-pressure control valve. This valve should be set at a pressure below that of the liquid cylinder relief-valve setting, thus ensuring that any excess pressure is safely vented.

6.75 Where installed in buildings, generous ventilation should be provided by means of fully-louvred access doors to the outside. The appropriate calculation must be made to ensure adequate ventilation, especially during the filling of the vessels, when they may be venting freely to atmosphere inside the manifold room.

6.76 A P&ID of the plant should be displayed clearly to indicate the appropriate valves necessary to operate the plant safely. The Authorised Person (MGPS) and others in the hospital who may work with the VIE system should be made aware of its general operating principles by the medical gas supplier. (Figure 21 shows a typical liquid cylinder manifold installation with cylinder backup.)

### *Compressed gas cylinder manifolds*

6.77 The simplest supply system to provide medical oxygen to a hospital pipeline system utilises compressed gas cylinders, connected together on an auto-changeover manifold. As the demand increases, the number of cylinders fitted to the manifold can be increased to meet the hospital's requirements. The secondary supply for this system will usually be a manual changeover compressed gas cylinder manifold, which comes on line automatically (via a non-return valve) in the event of primary manifold failure.

6.78 For a full description of manifold supply systems, see Chapter 5.

### *Secondary supply systems*

6.79 Where the primary supply system is a VIE, the secondary supply system can be either:

- another VIE; or
- a liquid cylinder manifold; or
- a fully-automatic compressed gas cylinder manifold.

6.80 Where the primary supply system is a liquid cylinder system, the secondary supply system should be a fully automatic changeover gas cylinder manifold that comes into operation automatically.

6.81 There should be a system of backflow prevention to protect either system venting through the other in the event of a single fault in either system.

## Figure 21 Typical liquid cylinder manifold installation with cylinder backup

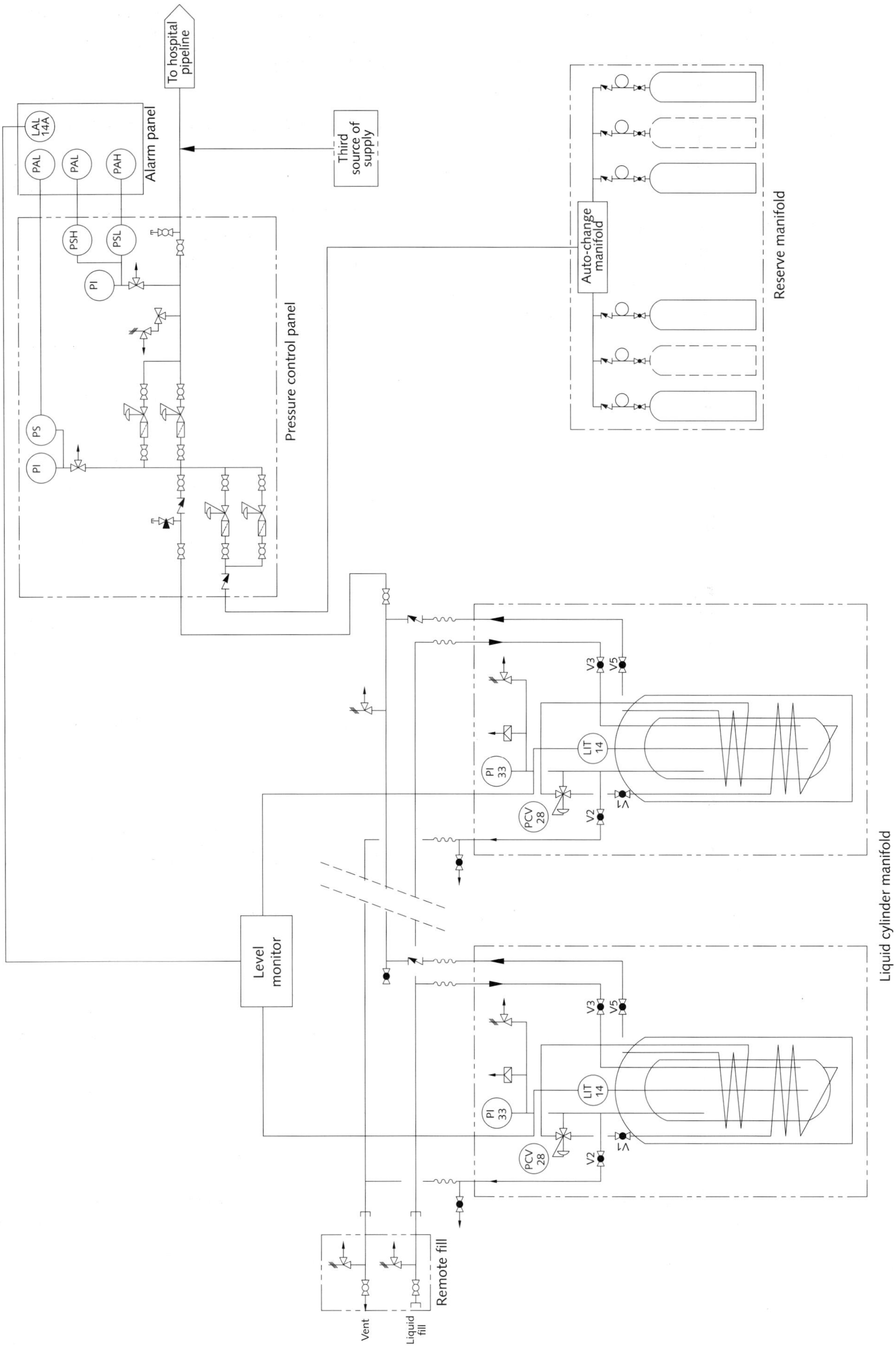

6.82 Where the secondary supply is fed from compressed gas cylinders, the size of the changeover manifold and the number of cylinders stored on site should be based on the gas supplier's ability to guarantee a delivery service within a defined period.

6.83 Where a liquid oxygen system is used for the secondary supply, the system design should allow any liquid oxygen that has boiled off to be fed to the pipeline system to utilise product.

6.84 Where the feed from the VIE compound to the hospital extends a long distance, or is exposed to potential mechanical damage, particular importance should be given to siting the secondary supply system remotely from the main VIE compound with a separate supply to the hospital pipeline system.

6.85 Where the secondary supply is sited remotely, consideration should also be given to the set point of the secondary supply output regulator to ensure that the pipeline pressure is maintained at a minimum level of 4.1 bar.

6.86 When the vessels are located on separate sites, a backflow prevention device must be fitted on each leg feeding into the pipeline system. This will prevent loss of product, either from the other vessel or from the hospital pipeline, in the event of failure of or damage to a VIE unit or its feed into the hospital pipeline. The backflow protection should be sited in a secure location where it is not liable to mechanical damage and be as close to the hospital curtilage as possible.

## Emergency supply provision

6.87 In the event of total plant and/or main pipeline failure, an emergency supply of oxygen should be available for patient use.

6.88 The emergency supply system should be activated automatically when the primary and secondary system is empty or fails to supply or when the hospital pipeline pressure falls below 3.8 bar. It must have the provision to automatically prevent the backflow of medical oxygen into the remainder of the pipeline system should the pipeline fail upstream of the connection.

6.89 A variety of sources are available for the provision of emergency oxygen, and these are detailed in paragraphs 6.96–6.105.

6.90 Under most conditions, compressed gas cylinders are the appropriate method of providing an emergency supply source.

6.91 The size and design of the emergency supply system should allow for cylinders to be changed whilst in operation.

6.92 Consideration should also be given to the set point of the regulator of any emergency supply system, to ensure that the pipeline pressure remains above the minimum level of 3.8 bar.

## Emergency inlet ports

6.93 Emergency inlet ports are covered in Chapter 13. In some instances, installation of a fixed manifold system will obviate the need for fitting an emergency inlet port.

6.94 In smaller installations, fitting an emergency inlet port may be dispensed with if the risk analysis indicates that adequate supplies can be maintained via the NIST connectors of AVSUs supplying critical care areas.

6.95 When planning emergency provision for a complete system, vulnerability of the primary and secondary supplies will be a critical factor in determining both the type and the means of supply.

## Fixed automatic/manual manifold systems

6.96 Where two VIE units on the same plinth are in use, the emergency supply system should comprise a fully automatic cylinder manifold permanently connected to (one of) the main oxygen riser(s) in the hospital, or directly into a ring-main system. It must be able to feed a riser automatically (without back-feeding to any damaged upstream section) on failure of both primary and secondary plant, or the MGPS upstream of the entry into the hospital. Such an arrangement is particularly suited to situations in which the main feed from the VIE installation to the hospital pipeline is vulnerable to mechanical damage, for example when buried under a roadway. The location and size of the manifold should be determined by the risk assessment and according to the dependency of the patients.

6.97 When two separately sited VIE units are used to provide the hospital supply, the need for emergency manifold provision should be assessed against the likelihood of failure of both VIE systems and their respective feeds into the hospital pipeline.

6.98 Where it can be shown that one or both units are fed into the hospital pipeline in a manner such that the probability of disruption of one or both of the feeds is negligible, the option to waive the fitting of further (manifold) supplies can be considered.

## Local manifold provision (critical care areas)

6.99 To offer additional protection against the possibility of a pipeline failure within the hospital, further (manual or automatic) manifolds can be permanently connected, via non-return valves, downstream of AVSUs controlling high-dependency areas. Such units should come on line automatically on failure of the main supply to an AVSU. A further non-return valve must also be added upstream of the AVSU to prevent back-feeding into a failed main supply system.

6.100 The positioning of these manifolds is very important to ensure that the critical supply/high-dependency areas defined in the risk assessment process have adequate stocks of medical oxygen available in the event of a system failure. However, the risk analysis for the complete system may indicate that the probability of use of such a manifold system is negligible, or that the circumstances causing the system failure would in any event require the evacuation of the area.

6.101 Availability of accommodation, staff and manual handling issues would also need to be considered during the risk assessment process. Where space limitations prevent the installation of such manifolds, the implications of providing discrete cylinder/regulator combinations must be considered.

## Gas feed via an AVSU (or terminal unit)

6.102 Oxygen supply to the downstream side of an AVSU (with the valve closed) may be achieved using an "emergency supply kit" consisting of two cylinder regulators and associated supply hoses with gas-specific connectors.

6.103 Such an arrangement may be used to support high-dependency departments, albeit the unit will usually be of limited capacity by comparison to a fixed automatic manifold system.

6.104 Storage, maintenance, testing, security and deployment arrangements for the emergency supply kits must be documented in the MGPS operational policy.

## Discrete cylinder supplies

6.105 For non-critical care areas where there are no high-dependency patients, it may be appropriate to use individual cylinders as the reserve source of supply. Cylinders fitted with integral valves and having a product-specific terminal unit outlet are suitable for this purpose. The difficulties associated with storing, testing, maintaining, distributing and connecting large numbers of such equipment must not be underestimated. (Such protocols should be referenced in the MGPS operational policy.)

## Alarm systems

6.106 Installations of the following type should be fitted with alarm systems to provide visual and audible warnings for the following conditions. For:

a. dual VIE vessels feeding into a single control panel; or

b. a single VIE vessel supported by a liquid cylinder secondary supply; or

c. a single VIE vessel supported by a fully automatic compressed gas cylinder manifold,

the following displays should be presented at the plant and in a 24-hour-staffed position.

| Status/fault condition | Indication | Legend |
|---|---|---|
| Normal operation<br>System available for use | Green | Normal |
| Primary supply system's operational stock empty<br>Primary supply system's reserve stock in use | Yellow | Refill liquid |
| Primary supply system's reserve stock empty<br>Secondary supply system in use | Yellow | Refill liquid immediately |
| Secondary supply system empty<br>Emergency system in use | Yellow | Emergency supply in use |
| Pipeline pressure high or low | Red | Pressure fault |

6.107 For:

a. two discrete VIE vessels feeding into separate parts of the pipeline system; or

b. a single VIE vessel supported by a liquid cylinder secondary supply that feeds separate parts of the pipeline system; or

c. a single VIE vessel supported by a fully-automatic compressed gas cylinder manifold secondary supply that feeds separate parts of the pipeline system, the following displays should be presented at the plant and in a 24-hour-staffed position.

| At the primary vessel and a 24-hour-manned position | | |
|---|---|---|
| **Status/fault condition** | **Indication** | **Legend** |
| Normal operation<br>System available for use | Green | Normal |
| Primary supply system's operational stock empty<br>Primary supply system reserve stock in use | Yellow | Refill liquid |
| Primary supply system's reserve stock empty<br>Secondary supply system in use | Yellow | Refill liquid immediately |
| Secondary supply system low | Yellow | Secondary stock low |
| Pipeline pressure high or low | Red | Pressure fault |

6.108 When the primary system operational stock has been exhausted and the vessel contents reach the reserve stock level, the first alarm condition will be indicated by a yellow alarm and the legend "refill liquid" illuminated.

6.109 When the primary system reserve stock is empty and the secondary supply system is in operation, the second alarm condition will be indicated by a yellow alarm and the legend "refill liquid immediately" illuminated. This alarm condition continues until the primary supply system is refilled.

6.110 When the secondary supply system is low, the third alarm condition will be indicated by a yellow alarm and the legend "secondary stock low" illuminated. This alarm condition continues until the secondary supply system is refilled or the cylinders are replaced.

6.111 Should the primary supply of medical oxygen to the hospital pipeline fail due to lack of contents or mechanical failure of any of the components, or should a serious leak occur, the pipeline pressure will fall. When the plant output pipeline pressure falls below 3.75 bar, the condition will be indicated by the "pressure fault" alarm.

6.112 If the regulator controlling the pipeline pressure should fail "open", the pipeline pressure will rise. This condition will be indicated by the "pressure fault" alarm when the pressure rises above 4.9 bar.

| At the secondary vessel/liquid cylinder supply/cylinder manifold and a 24-hour-staffed position | | |
|---|---|---|
| **Status/fault condition** | **Indication** | **Legend** |
| Normal operation<br>System available for use | Green | Normal |
| Secondary supply system's operational stock empty<br>Secondary supply system's reserve stock in use | Yellow | Refill liquid/ change cylinders |
| Secondary supply system's reserve stock empty<br>Emergency supply system in use | Yellow | Emergency supply in use |
| Pipeline pressure high or low | Red | Pressure fault |

6.113 When the secondary system operational stock has been exhausted and the vessel contents reach the secondary reserve stock level, the first alarm condition will be indicated by a yellow alarm and the legend "refill liquid" illuminated. This alarm condition continues until the secondary supply system is refilled.

6.114 When the secondary supply system is empty, the second alarm condition will be indicated by a yellow alarm and the legend "emergency supply in use" illuminated. This alarm condition continues until the secondary supply system is refilled or the secondary supply manifold cylinders are replaced.

6.115 Should the secondary supply of medical oxygen to the hospital pipeline fail due to lack of contents or mechanical failure of any of the components or should a serious leak occur, the pipeline pressure will fall. When the plant output pipeline pressure falls below 3.75 bar, the condition will be indicated by the "pressure fault" alarm.

6.116 With a correctly installed and commissioned emergency system, the hospital pipeline pressure will be maintained downstream of the primary and secondary supply connections (non-return valves) at a minimum pressure of 3.9 bar. Pressure on the plant side of the non-return valves will, however, remain below 3.75 bar until the plant is refilled, or any fault remedied. Therefore, the plant and 24-hour-monitored alarms will continue to indicate both "pressure fault" and "emergency supply system in use" until the vessel is refilled or any fault rectified.

6.117 If the regulator controlling the pipeline pressure should fail in the "open" position, the pipeline pressure will rise. This condition will be indicated by the "pressure fault" alarm when the pressure rises above 4.9 bar.

**6.118** Where the emergency supply is installed on individual zones of the pipeline system, the "emergency supply in use" alarm must be displayed within the pipeline zone area. A separate "emergency supply low" alarm should also be installed on each zone.

**6.119** Where more than one VIE is used and the operational and reserve stock is distributed between multiple vessels, a lit "normal" display indicates that the vessel is available for use.

**6.120** In the event that the primary (or secondary) vessel should become empty (or suffer from any other fault condition), the "normal" display should be extinguished, indicating that the vessel is not available for use.

**6.121** Alarm conditions should be transmitted to the central alarm system.

**6.122** Where relays are used, they should be energised relays that de-energise under fault conditions, with contacts having a minimum rating of 50 V dc and 50 mA. Alternatively, volt-free, normally closed contacts rated at 50 V dc and 50 mA should be provided for transmission of the conditions to the alarm system.

**6.123** Typical alarm trigger points are shown in Figures 12–17.

## Determining system size through risk assessment

### *Introduction*

**6.124** The 1997 edition of Health Technical Memorandum 2022 defined a (fixed) VIE primary vessel capacity of 14 days' oxygen supply but did not define capacity of a liquid cylinder system. This section addresses the risk factors associated with the supply of oxygen on a hospital site and, with the aid of defined risk criteria, offers guidance on the sizing of VIE, liquid cylinder and compressed gas cylinder manifold installations for any specified location.

**6.125** The risk assessment should take into account all issues concerning the safety and continuity of the medical oxygen supply. It is suggested that identified risk factors and criteria be evaluated using both qualitative and quantitative measures, and that all results be recorded in a logical manner that will support the decisions being made. The record of the risk assessment will also act as a reference document when the system is reviewed.

**6.126** Additional local factors and requirements identified by the project team will also need to be considered when carrying out the risk assessment to take account of site-specific issues concerning how the product is stored, distributed and used.

**6.127** Any risk control procedures identified by the risk assessment process which are designed to minimise any identified risks must be recorded and incorporated into the relevant hospital standard operating procedures (SOP) or work instructions (WI).

**6.128** When sizing the vessels and cylinder manifolds to provide adequate storage of medical oxygen on site, the stock should be distributed between the three sources of supply as defined in BS EN 737-3: 2000; that is, the medical oxygen supply system should normally consist of:

- a primary supply;
- a secondary supply;
- a emergency supply.

**6.129** The capacity of the primary and secondary supply system will consist of:

- operational stock;
- reserve stock.

**6.130** The operational stock is the volume of product that the gas supplier uses to manage deliveries to the hospital, and its exhaustion signals the point at which the vessel should be refilled under normal conditions.

**6.131** The reserve stock is the volume of product that is used to provide additional stock, to take account of fluctuations in demand, or when the supplier fails to make a scheduled delivery.

**6.132** The system should be designed so that the primary and secondary supply system stocks are kept separate from each other. Under no circumstances can the primary supply system operational stock be stored in the secondary supply system vessel.

**6.133** However, where it is not possible to install a single large VIE vessel for the primary supply (such as where planning permission restrictions prevent the use of a single large vessel), it may be appropriate to hold all or some of the primary supply system reserve stock in the secondary supply vessel. Under these circumstances the primary supply vessel should retain a minimum level when changing over to the secondary supply system. The volume retained in the primary supply vessel should equate to the secondary supply system reserve stock. This should provide adequate stock on site to enable the gas supplier to resupply product to the primary vessel in the event of failure of the secondary supply. This level should be determined by the risk assessment process but should be at least one day's usable stock.

## *Review of risk assessment*

6.134 The documented risk assessment should be reviewed after the installation is complete, prior to commissioning, to assess whether any parameters or circumstances have changed since the initial assessment. The risk assessment must also be reviewed at least annually (or when there is any significant change to the medical oxygen supply system or usage pattern) to ensure that the details are current. At this review, all changes should be considered that might have an effect on the safety of the system or the security of supply.

## *Sizing plant – general*

### *VIE installations*

6.135 The operational and reserve stock for each supply system should normally be held in the same vessel. Where planning restrictions prevent the use of a single large vessel on site, it may be appropriate to utilise multiple vessels to provide adequate stocks on site.

6.136 When sizing VIE systems for the primary or secondary supply, the vessel size will be determined by adding the operational and reserve stock together and allowing for the level of unusable stock left in the vessel when the designed flow rate cannot be maintained.

### *Liquid cylinder installations*

6.137 For liquid cylinder installations, the primary system should be made up of a number of liquid cylinders connected together by a manifold. The secondary system will comprise an automatic or manual compressed gas cylinder manifold system.

6.138 Each liquid cylinder will have a maximum design flow rate for continuous use. The number of liquid cylinders required for an installation may be governed by either the maximum storage capacity required on site or the flow rate required to meet the hospital's maximum demand.

6.139 When determining the number and size of liquid cylinders required for either a primary or a secondary supply to a VIE, an allowance has to be made for the unusable capacity of each cylinder when connected to the manifold system.

### *Compressed gas cylinder manifold systems*

6.140 For auto-changeover cylinder manifolds, one bank of cylinders should be considered as the operational stock and the other bank as the reserve stock. The size of each source of supply should be determined by considering the operational and reserve stock requirements for each source.

6.141 The secondary supply will normally comprise a manually operated manifold system, connected such that it will come on line automatically (via a non-return valve) in the event of primary supply failure.

6.142 For sizing compressed gas cylinder systems, the size of the manifold will normally be determined by the ability of the hospital to provide adequately trained staff to change over cylinders quickly enough to meet the demand.

6.143 The number of cylinders required to support the manifold can be determined by dividing the relevant stock figure by the usable volume of each cylinder (that is, the volume at full cylinder pressure less the volume at the pressure of the cylinder when the manifold changes over).

## *The risk assessment process*

### *Risk assessment for management responsibilities*

6.144 The risk assessment criteria, when considering management responsibilities for the medical liquid oxygen system, need to include the following:

- the need to document and agree responsibilities for the monitoring of the medical liquid oxygen VIE, and the need to establish a back-up procedure with the gas supplier to ensure that adequate stocks will be maintained in the event of a failure of the fitted telemetry system;
- the hospital should set up procedures to ensure that the VIE system is monitored at regular intervals for any deviation from normal operation (such as safety valves lifting, major leaks, or failure of either the telemetry or alarm system);
- the implications of any decisions to not fit telemetry or to utilise a vessel, or vessels, that do not provide adequate operational and

reserve stocks. These decisions should be taken at an appropriate level of management, should be documented, and their implications should be considered as part of the risk assessment;

- consideration of the resources needed to maintain adequate supplies of medical oxygen either under normal, or emergency, conditions. When evaluating these requirements, consideration should be given to the risks that the trust would face in the event of supply failure causing disruption of clinical services;
- consideration of the operational management consequences of using different suppliers to supply medical oxygen to different supply systems supporting the same pipeline installation. Any contracts involving different suppliers should clearly state the obligations and limitations of liabilities.

6.145 Where manifolds are used as either the secondary or emergency supply, adequately trained staff should be available, whenever required, to ensure continuity of supply. Consideration also needs to be given to the manual handling issues concerned with changing cylinders on the manifold and arrangements to store adequate stocks to meet demands.

6.146 Consideration needs to be given to the type of clinical activities carried out in each area of the hospital and the ability to provide emergency back-up to individual areas used for critical care, or within high-dependency units.

*Initial risk assessment for siting of plant*

6.147 The initial risk assessment should consider the requirements to ensure a continuous supply of medical gas to the patient.

6.148 The initial risk assessment criteria related to the complete installation should include:

- the size and location of each source of supply (for example the volume held as operational and reserve stock for each source of supply, located on one site or two independent sites);
- the associated risks with siting tanks at either the same or separate locations (for example physical space availability, accessibility for delivery and maintenance requirements, accessibility to the pipeline system [to tie-in points etc], alarm systems and cabling, pipeline routing and protection);
- the need to site the reserve sources of supply local to the point of use to protect against pipeline failure where high-dependency patients are located;
- safety requirements for the storage of oxygen on site, including compliance with the safety distances specified in BCGA CP19;
- the location and extent of the medical oxygen pipeline system;
- the vulnerability of the hospital pipeline to mechanical damage and whether underground sections of the pipeline system comply with the requirements of this Health Technical Memorandum; and whether the pipeline is capable of being inspected throughout its entire length or pressure tested (whilst maintaining the supply), or otherwise can be tested;
- the space available for the liquid oxygen installation, or cylinder manifold, and the available access for the delivery vehicle;
- the vulnerability of the site to external damage;
- the possibility of interference with the supply system or other security issues.

*Risk assessment for sizing of operational stock*

6.149 The risk assessment criteria for the sizing of the operational stock should include:

- the average daily demand at the end of the contract period. Any changes to the predicted growth of demand will need to be considered, and changes made to the vessel size or delivery frequency at the appropriate time within the contract period. It may be beneficial to set a daily demand rate at which changes to vessel size or delivery frequency will be considered;
- a review of vehicular access to the VIE, timing of the deliveries, any restrictions due to local planning requirements, and the effect of these factors on the delivery frequency;
- an environmental impact assessment.

*Risk assessment for sizing of reserve stock*

6.150 The risk assessment criteria for the sizing of the reserve stock should include:

- the average daily demand at the end of the contract period. Any changes to the predicted growth of demand will need to be considered, and changes made to the vessel size or delivery frequency at the appropriate time within the contract period. It may be beneficial to set a daily demand rate at which changes to vessel size or delivery frequency will be considered;

- the delivery frequency guaranteed by the gas supplier that can be provided at short notice should the primary supply system fail;
- the minimum response time from when the primary supply system fails to when the delivery vehicle could be on site to refill the secondary supply VIE, or to provide replacement compressed gas cylinders for the manifold.

*Risk assessment for the provision of emergency supply systems*

6.151 The risk assessment criteria concerning emergency supply systems should include:

- the need for installation of independent emergency supplies to zones on the medical gas pipeline supplying critical care areas or wards or departments that are remote or vulnerable to interruption;
- the positioning of the manifold to ensure ease of changeover of cylinders with respect to access and manual handling issues;
- the storage of cylinders associated with the emergency manifold to ensure compliance with the appropriate codes of practice and local hospital requirements;
- training requirements for both the relevant clinical and operational staff to ensure correct operation of the emergency supply system.

## Stock calculations

*Calculation of operational stock for primary and secondary supplies*

6.152 The capacity of the operational stock of primary and secondary supply systems should be agreed with the gas supplier and based on the following parameters:

- the current average medical oxygen daily demand, plus any natural growth over the contract period;
- any additional planned growth (above any natural growth) in the usage pattern within the contract period;
- the agreed delivery frequency.

6.153 The current average daily demand can be calculated by dividing the current annual consumption by 365 days.

6.154 The operational stock should be based on an average daily demand predicted for the end of the contract period calculated by:

Average daily demand = Current daily demand + Planned growth + Natural growth.

6.155 The operational stock is calculated as:

Operational stock = Average daily demand x Agreed delivery period.

6.156 If there is significant growth in average daily demand within the contract period, either the vessel should be resized or the agreed delivery frequency should be reviewed to reduce the delivery period and maintain the operational stock level.

6.157 The delivery period for the primary supply will be based on the gas supplier's normal delivery frequency.

6.158 The delivery period for the secondary supply will be based on emergency conditions when the primary supply is not available. Under these circumstances, special delivery response times must be agreed with the gas supplier.

6.159 The supply agreement should commit the supplier to manage the operational stock, based on an agreed delivery frequency and the minimum stock level to be maintained in the vessel.

*Calculation of primary reserve stock*

6.160 The table below provides a matrix for the calculation of primary reserve stock based upon distance from gas supplier and fitting of telemetry.

| Kilometres from gas supply depot | No telemetry (no of days' stock) | Telemetry fitted (no of days' stock) |
|---|---|---|
| Up to 75 | 5 | 3 |
| 75–150 | 6 | 4 |
| 150–300 | 7 | 5 |
| Over 300 | 8 | 6 |

*Calculation of secondary reserve stock*

6.161 The minimum level for reserve stock for the secondary supply should allow for circumstances in which the primary supply system is not available for use.

6.162 This secondary supply system reserve stock level will be dependent on:

- the proximity of the supplier's distribution depot;
- the response time that the gas supplier needs to make a delivery under these conditions;
- the delivery frequency that can be sustained under the conditions when the primary supply is unavailable for use.

*Calculation of capacity of emergency supply systems (cylinder manifolds)*

6.163 The number of cylinders stored locally to the emergency supply system manifold and the number of connections on the manifold(s) should be determined by risk assessment.

6.164 When determining these requirements, the risk assessment needs to consider:

- the maximum demand from the high-dependency patients who may be supplied from the pipeline zone that the emergency supply system protects;
- the maximum duration for which the emergency state is likely to last;
- the proximity of the supplier of the compressed cylinders to the hospital;
- the ability of the hospital to connect cylinders to the manifold.

6.165 Consideration needs to be given to the logistics of storing and handling the number of cylinders needed to provide adequate supplies until the primary/secondary supply systems or the hospital pipeline can be re-established.

## Oxygen concentrator installations (PSA plant)

### General

6.166 Oxygen concentrators or pressure swing adsorber (PSA) systems may be an alternative to the more traditional supply systems (the terms oxygen concentrator and PSA are interchangeable). Typical installations where PSA systems should be considered are those sites having no access to reliable liquid supplies, such as remote or off-shore locations, or where the safety criteria for a bulk liquid vessel cannot be met (for example, very restricted sites). Otherwise, PSA systems should only be installed when an investment appraisal shows them to be economical.

6.167 When installed, a PSA system will deliver product gas via the "oxygen" pipeline system.

6.168 Oxygen concentrators operate by adsorbing, under pressure, other gases in the atmosphere onto materials which have specific physico-chemical properties, thus freeing the oxygen which is stored and transmitting it for use. The adsorbents are known as artificial zeolites, more commonly referred to as molecular sieves. The sieve units are arranged in pairs, one adsorbing whilst the other regenerates. The waste product, essentially nitrogen, is discharged to atmosphere during regeneration of the adsorbents. In some systems, the use of vacuum to remove the nitrogen increases the efficiency of the regeneration/adsorption process. Regeneration requires the use of a small proportion of the product gas.

6.169 The PSA process has reached a high level of technical sophistication and is capable of producing oxygen with a concentration of about 95%. (For the UK the minimum level, below which the emergency/reserve manifold will come into operation, is 94%.) The remainder is mainly argon with some nitrogen. The highest concentration is not likely to exceed 97/98%, except when the emergency/reserve manifold is in use, when it will be 100% if these are from a gas supplier.

6.170 The major components of a PSA system and their layout are shown in Figure 22. The typical major components of the system are the compressors, receiver(s), dryers, molecular sieves, vacuum pumps, filters and regulators. Other components are identical to those used for medical air and vacuum plant, which are described fully in the appropriate sections. A suitable operating and indicating system is also required, as specified below. Package supply systems, which should be specified to meet the requirements given in this memorandum, are available from manufacturers.

### Siting

6.171 The plant should have all-round access for maintenance purposes, and allowance should be made for changing major components.

6.172 The siting of the plant should allow for adequate flows of air for three different purposes:

a. air intake to the compressors;

# Figure 22 Schematic diagram of a typical PSA installation

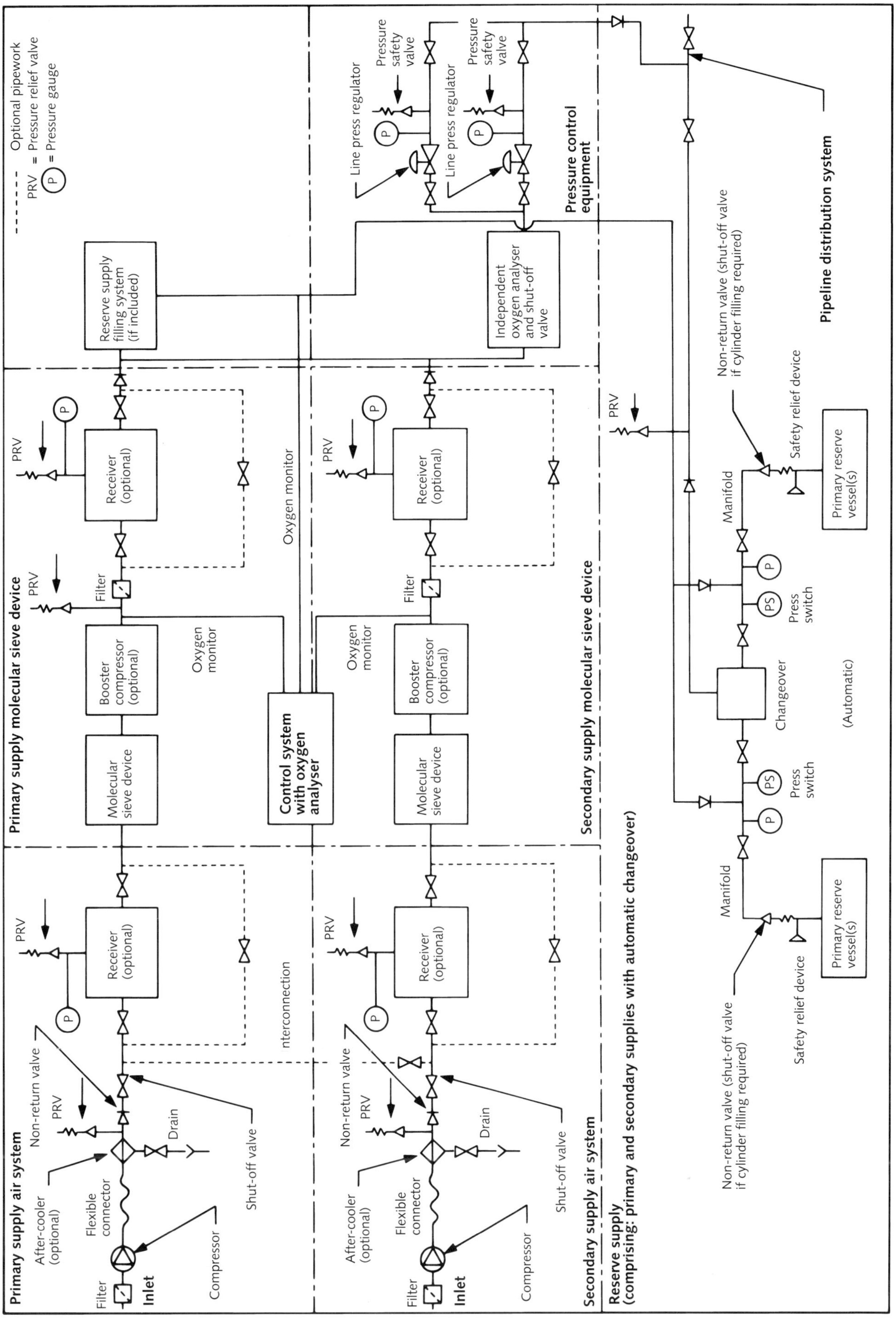

b. cooling of the compressed air by the after-coolers;

c. cooling of the compressors.

6.173 Each compressor may require ducting to ensure an adequate flow of cool air. The manufacturer should be consulted over the range of operating temperature for which the system is designed. In extreme circumstances, refrigeration of the cooling air may need to be provided.

6.174 Air-inlet filters should be fitted either to the compressor inlet or at a suitable point in any ductwork. The filters should comply with BS ISO 5011:2000 and be either dry medium filters or grade CA paper element filters.

## Plant configuration

6.175 The plant should comprise:

a. a duplex compressor – if more than two compressors are installed, the plant should provide the design flow with one compressor out of service;

b. duplexed air treatment/molecular sieve devices, that is, two sets of filters and a pair of molecular sieves (one adsorbing whilst the other regenerates) and one vacuum pump (if required by the manufacturer).

> **Note**
>
> All duplexed components should be capable of independent operation.

## Compressors and vacuum pumps

6.176 The compressors for the PSA systems may be any of the type recommended for compressed air systems. It is also possible to provide a combined medical air PSA plant. Generally, the compressed air requirement per litre of product gas is of the order 4:1; as a result the compressor plant will be on longer than that typically seen in hospitals.

6.177 A vacuum pump may be required as part of the system. The vacuum pump, if provided, is utilised during the adsorption/regeneration process. Vacuum pumps may be of any type as for the piped medical vacuum system. It will not generally be practicable to use water-sealed pumps or the medical vacuum plant.

## Compressor and vacuum pump noise

6.178 The noise level produced by the compressors will increase with the capacity of the supply system. The maximum free-field noise level for unsilenced compressed air plant, at 1 m from the plant, varies with the type and power of the plant but should not normally exceed the following values:

| Reciprocating | Screw | Vane | Power |
|---|---|---|---|
| 85 dBA | 76 dBA | 76 dBA | 7.5 kW |
| 89 dBA | 78 dBA | 76 dBA | 7.6–15 kW |
| 93 dBA | 80 dBA | 79 dBA | 15.1–22 kW |
| 97 dBA | 92 dBA | 90 dBA | 22.1–60 kW |

6.179 In noise-sensitive areas, an acoustic enclosure should be included in the purchase specification for all compressors. Such an enclosure should produce a reduction of at least 10 dBA in the free-field noise level at 1 m.

## Molecular sieves

6.180 Duplex molecular sieves should be provided in pairs to permit continuous generation of oxygen. One of the pairs of duplex sieves will be in the adsorbing stage, whilst the other regenerates.

## Dryers

6.181 Air dryers of the desiccant type are usually integrated within the molecular sieves and therefore do not regenerate independently. Refrigerant dryers may also be included.

## Oxygen monitoring system

6.182 The plant should include a calibrated paramagnetic oxygen monitoring system comprising oxygen analyser, oxygen concentration indicator, oxygen flow monitor and oxygen concentration/flow recorder. Connections for calibration cylinders should also be provided. In the event of the concentration falling below 94%, the monitoring system should isolate the PSA system from the pipeline distribution system so that the emergency/reserve manifold operates. Additionally, an independent monitoring system should be provided to isolate the plant when the concentration falls below 94%. The second system need not be provided with a flow indicator or recorder.

## Operating and indicating system

6.183 The operating and indicating system should perform the following functions, as appropriate:

a. overall plant control and indication;

b. individual compressor starting;

c. individual vacuum pump starting (where fitted);

d. control of dryers (where installed as separate component);

e. control of molecular sieves;

f. plant status monitoring and indication;

g. optional indication of the plant alarm status (this function may be considered to be part of the alarm system).

6.184 Provided that the individual compressor starters are housed in a separate compartment, these functions may be carried out by separate units or may be installed in a common panel and located on the plant or on the plantroom wall.

6.185 Control panels containing pneumatic components should have vents to permit release of pressure in the event of component failure. All functions and indicators should be appropriately identified and should have a design life of at least five years. The operating system should be capable of automatically restarting after reinstatement of the power supply.

6.186 All components of the PSA supply system should be connected to the essential electrical supply. The control system should ensure that compressors restart in sequence to avoid overloading the essential power supply.

## Plant control unit

6.187 The plant control unit should have a separate power supply for each compressor and vacuum pump, controlled by a separate sub-circuit. The design should be such that no single component failure in the control unit will result in loss of plant output.

6.188 The unit should allow either manual selection of duty/stand-by for each of the compressors or have an automatic sequence selection with a means for manual override. The unit should ensure that two or more compressors do not start simultaneously when power is applied.

6.189 A warning notice that complies with BS 5499-5: 2002 should be affixed which indicates the presence of low voltage.

6.190 A further warning notice indicating that the plant starts automatically should also be affixed near or on the plant.

6.191 Each compressor should have a selector switch which, when turned to the "on" position, allows the maximum and minimum pressure switches on the receiver to control the "on" and "off" loading of that compressor. An alternative "auto" position of the selector switch may allow automatic selection of the compressors.

## Plant control indication

6.192 There should be indicators for each compressor as follows:

a. green "mains supply on";

b. green "compressor called for", which indicates that the compressor motor is electrically energised;

c. an indicator of the pressure produced by the compressor.

## Compressor and vacuum starter units

6.193 There should be individual starter units for each compressor and vacuum pump, which should include the features recommended for medical air compressor plants and vacuum plants respectively.

## Molecular sieve control unit

6.194 The molecular sieve control unit may be mounted on the molecular sieve columns or may be located with the plant control unit. There should be separate power supplies for the "duty" and "stand-by" sieve assemblies, taken from the same phase.

6.195 The vacuum pump, if provided, forms part of the molecular sieve system.

6.196 The molecular sieve control unit should contain the following:

a. a duty selector switch;

b. an on/auto selector switch;

c. individually-fused, separate cycling systems for each sieve pair;

d. a system to control regeneration of the sieves in relation to pipeline demand;

e. oxygen concentration, dryness and pressure sensors;

f. an automatic changeover to the stand-by molecular sieve system, in the event of failure of the duty unit by oxygen concentration, dryness or pressure. This requires:

(i) electrical and pneumatic isolation of the "duty" sub-assembly so that it is taken off-stream;

(ii) electrical and pneumatic energisation of the "stand-by" sub-assembly so that it is brought on-stream;

(iii) activation of the appropriate fault indicator and associated volt-free contacts;

(iv) the sub-assembly to remain in this mode of operation until the fault has been rectified;

g. green function indicators for each dryer sub-assembly to indicate:

(i) molecular sieve 1 selected;

(ii) molecular sieve 2 selected;

(iii) selected molecular sieve – "normal";

(iv) selected molecular sieve – "failed" (this fault indicator should remain until manually reset by means of a reset button);

h. a fail-safe system that, on failure of the power supply, causes the closure of all inlet, outlet, exhaust and purge valves.

## Plant status monitoring

6.197 A monitoring system must be provided to detect the following faults in the air compressor system:

a. plant faults (for each compressor):

(i) control circuit failed;

(ii) overload tripped;

(iii) after-cooler temperature high;

(iv) compressor temperature high;

(v) compressor run-up time too long;

(vi) activation of other safety devices supplied by the manufacturers;

b. plant faults (for each molecular sieve unit):

(i) control circuit failed;

(ii) "vacuum pump called for";

(iii) overload tripped;

(iv) activation of any of the safety devices supplied by the manufacturer;

(v) oxygen concentration failure;

(vi) pressure fault;

c. plant emergency:

(i) oxygen concentration failed at below 94% concentration;

(ii) receiver pressure 0.5 bar (gauge pressure) below the stand-by cut in pressure;

(iii) dryness above 67 ppm (–46°C at atmospheric pressure);

d. pressure fault (cylinder reserve):

(i) pressure in either bank below 50% (of normal cylinder pressure);

e. pressure fault (pipeline):

(i) low pipeline pressure;

(ii) high pipeline pressure.

## Plant status indicator unit

6.198 In addition to the plant control indication, there should be a plant status indicator panel, which may be mounted on the plantroom wall or adjacent to either the compressor starter unit or the plant control unit. It should have a warning notice that complies with BS 5499-5:2002 to indicate the presence of low voltage.

6.199 There should be indicators for each compressor to show the following conditions:

a. green "mains supply on";

b. yellow "control circuit failed";

c. yellow "overload tripped";

d. yellow "after-cooler temperature high";

e. yellow "compressor temperature high";

f. yellow for each individual safety device provided by the manufacturers;

g. yellow "compressor failure".

6.200 There should be indicators for each molecular sieve dryer system to show the following:

a. green "mains supply on";

b. yellow "oxygen concentration fault";

c. yellow "pressure fault";

d. yellow “dryness fault”.

(When the stand-by dryer is in operation, conditions (b) and (c) in paragraph 6.199 should be transmitted as a plant emergency either to the alarm system or to the plant alarm signal status unit.)

## Alarm signal status unit

6.201 An alarm signal status unit should be provided as part of the control system. It should display the following conditions:

| Indication | Legends |
|---|---|
| a. Green “normal” | Normal conditions |
| b. Yellow “plant fault” | Conditions (b)–(f) (see paragraph 6.199) |
| c. Yellow “plant emergency” | Conditions (b) or (c), or condition (g) (see paragraph 6.199) |
| d. Yellow “emergency supply low” | Emergency supply bank(s) low (<50%) |
| e. Red “pipeline pressure fault” | Pressure fault |
| f. Red “pipeline concentration below 94% $O_2$ | Oxygen concentration fault |
| Conditions (b) to (e) should be transmitted to the central alarm system. | |

6.202 Where relays are used, they should be normally energised relays, which de-energise under fault conditions, with contacts having a minimum rating of 50 V dc and 50 mA.

6.203 Alternatively, volt-free, normally closed contacts rated at 5 V dc and 50 mA should be provided for transmission of conditions (b) to (e) to the alarm system.

6.204 The panel can be incorporated into the plant indicator unit, or be a separate unit within the plantroom. If mounted separately, the cabling should be monitored for open/short circuit. In the event of such a cabling fault, a red “system fault” lamp should be illuminated on the alarm signal status unit, together with the appropriate alarm condition.

6.205 The alarm signal status unit should be supplied from all individual plant control units, or from a separate common supply.

## Plant management

6.206 Connections should be provided that allow monitoring (but not control) of the plant operation. For example:

- compressor – “on”, “off”, “on-load”, “unloaded”;
- molecular sieves – “on” or “off”.

6.207 These connections should be used to provide input to the hospital energy management and building management systems.

# 7 Medical compressed air systems

## General

7.0 Medical compressed air can be derived from compressor systems or by mixing gaseous oxygen and nitrogen from cryogenic liquid supply sources; air produced by this latter method is referred to as synthetic air.

## Compressor systems

7.1 Medical and surgical air can be provided from a single combined system or from separate plants. The choice ultimately depends on the relative consumption.

7.2 In the case of surgical air, consumption is at a high flow at a high pressure (up to 350 L/min) but for relatively short periods of time (minutes); also, very small numbers of terminal units are in simultaneous use, typically fewer than five. Air for respirable purposes, however, is used at much lower flows (typically less than 80 L/min) but, particularly in the case of patient ventilation, use can be continuous. Moreover, in the case of medical air there are considerably greater numbers of terminal units in use simultaneously. The installation of separate plants therefore can result in lower running costs. Requirements for separate surgical air systems are given in Chapter 8.

7.3 Some plant configurations are shown in Figures 23–26:

- Figure 23 shows a duplex system with fully automatic emergency reserve manifold (the manifold is located in separate accommodation);
- Figure 24 shows a triplex system, each of the compressors to be capable of supplying the total design flow. It should be assumed that the third means of supply, an automatic manifold(s), will supply the distribution system at some point close to or within the building;
- Figure 25 shows a quadruplex system, each compressor being capable of supplying 50% of the design flow. It should be assumed that the third means of supply, an automatic manifold(s), will supply the distribution system at some point close to or within the building;
- Figure 26 shows a combined medical and surgical air system with automatic emergency reserve manifolds (located in separate accommodation).

7.4 Other plant layouts are possible depending upon the specific design requirements.

## Quality

7.5 The European Pharmacopoeia (Ph. Eur.) specifies maximum impurity levels for carbon monoxide. It may be necessary to make provision to control the levels of contaminants and to monitor the supply to ensure conformance with the specification. European Commission directive 2001/83/EC specifies that medicinal products should be manufactured to the approved standard.

## Siting

7.6 The plant should have all-round access for maintenance purposes, and allowance should be made for changing major components.

7.7 The siting of the plant should allow for adequate flows of air for three different purposes:

a. air intake to the compressors;

b. cooling of the compressed air by the after-coolers;

c. cooling of the compressors.

## Compressor noise

7.8 The noise level produced by the compressors will increase with the capacity of the supply system. The maximum free-field noise level for unsilenced compressed air plant, at 1 m from the plant, varies with the type and power of the plant but should not normally exceed the following values:

## Figure 23 Typical duplex medical air 400 kPa plant and energy reserve manifold (reproduced by kind permission of MEDÆS)

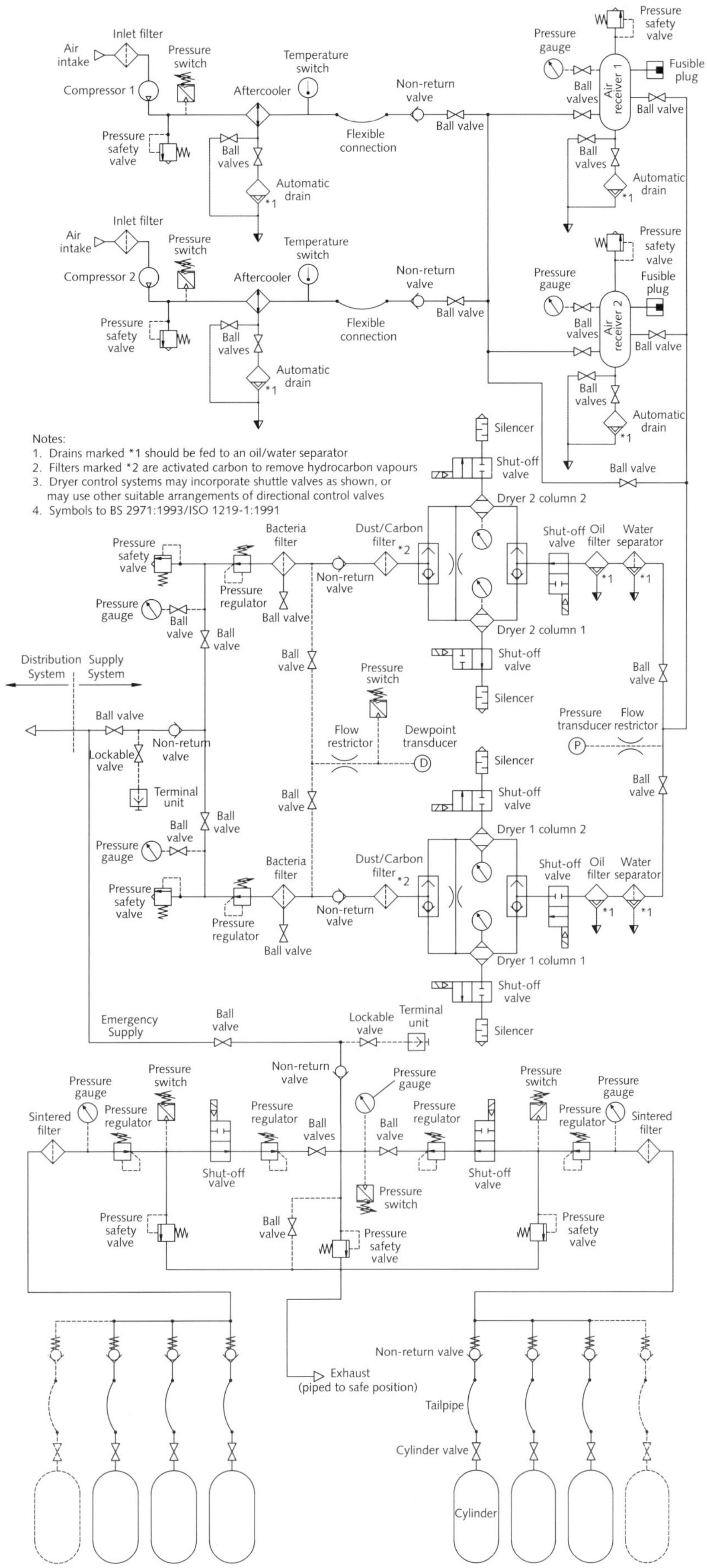

## Figure 24 Typical triplex medical air 400 kPa system (reproduced by kind permission of MEDÆS)

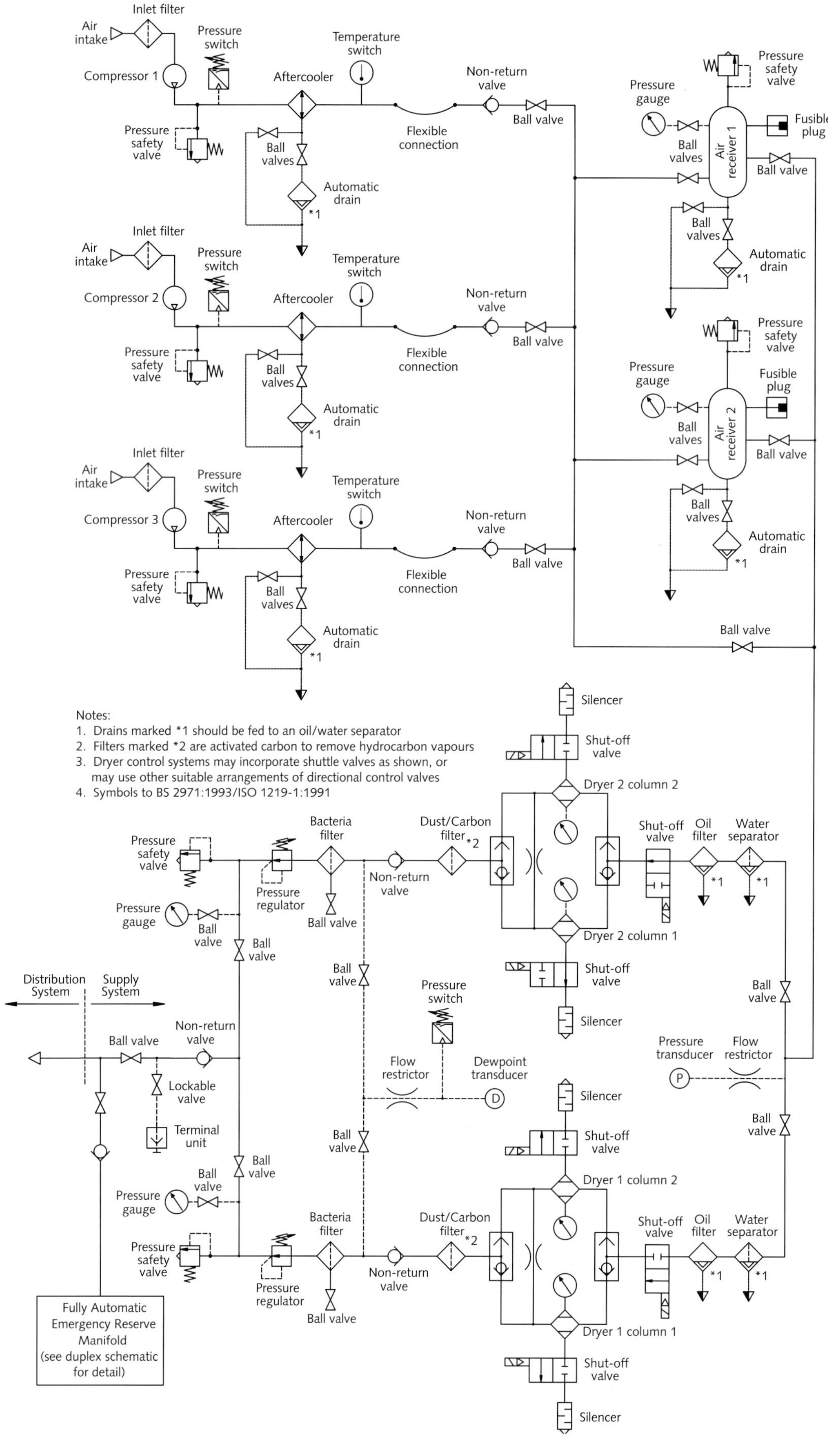

## Figure 25 Typical quadruplex medical air 400 kPa plant (reproduced by kind permission of MEDÆS)

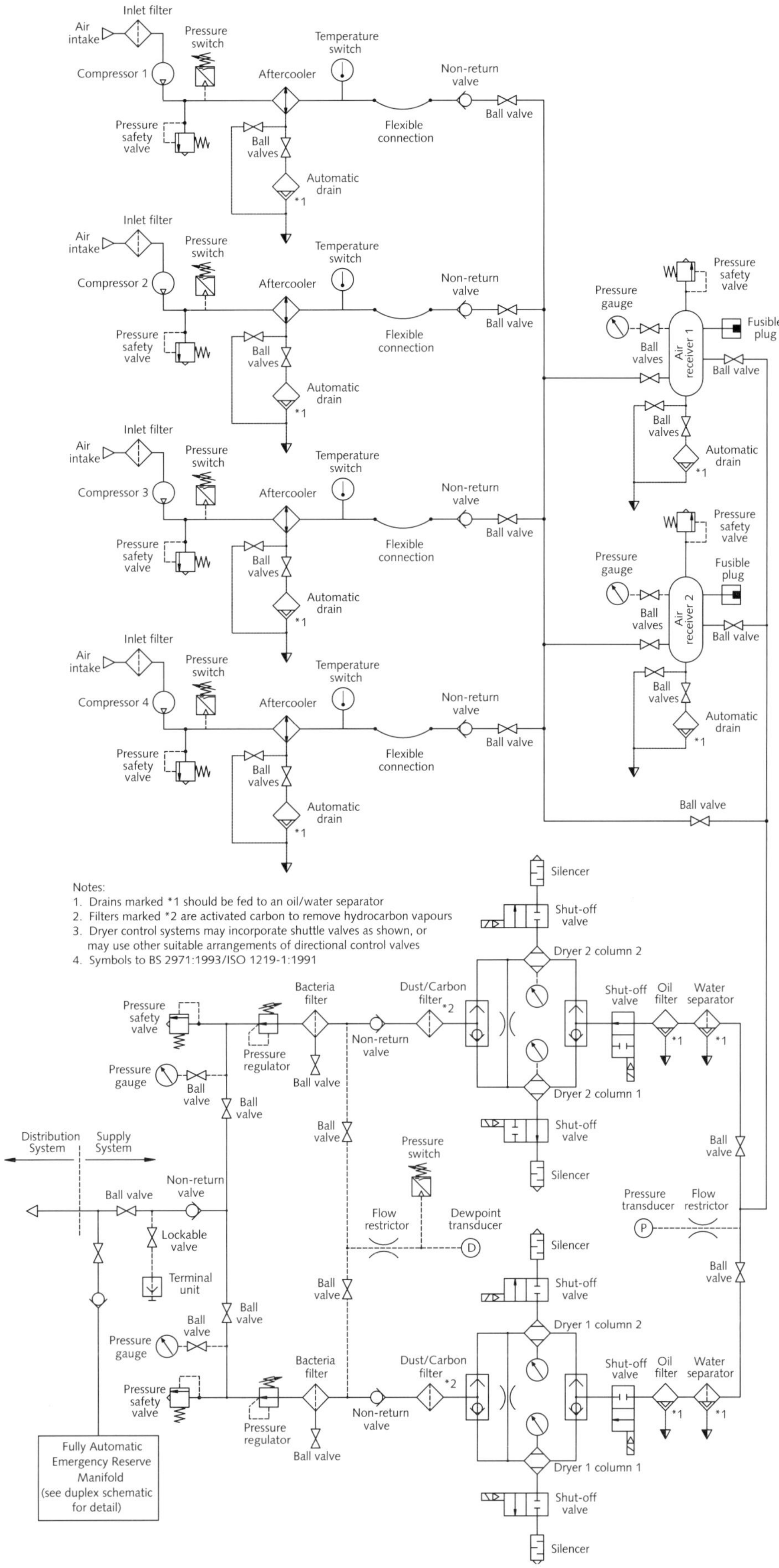

## Figure 26 Typical duplex combined medical and surgical plant with emergency reserve manifolds (reproduced by kind permission of MEDÆS)

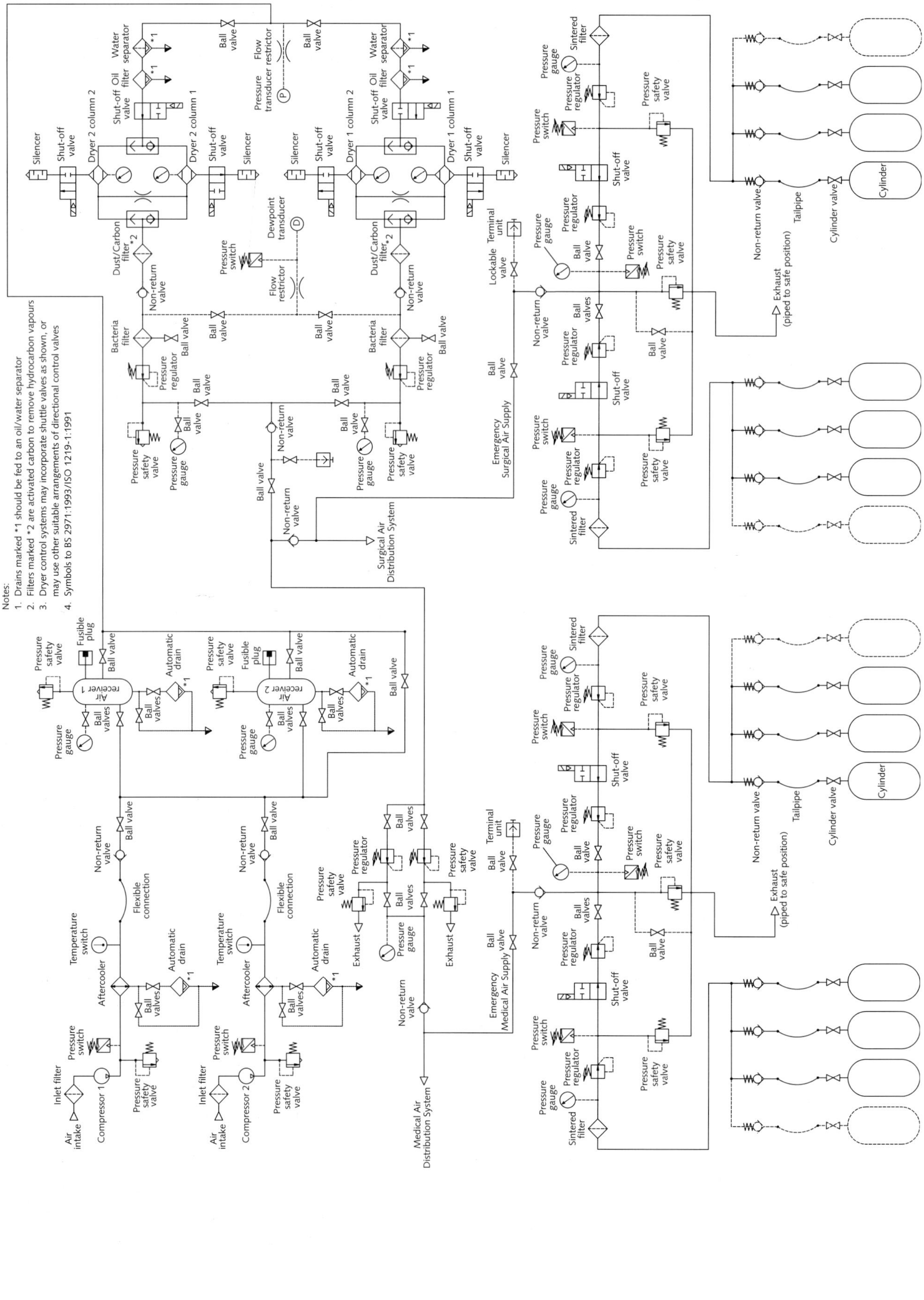


Notes:
1. Drains marked *1 should be fed to an oil/water separator
2. Filters marked *2 are activated carbon to remove hydrocarbon vapours
3. Dryer control systems may incorporate shuttle valves as shown, or may use other suitable arrangements of directional control valves
4. Symbols to BS 2971:1993/ISO 1219-1:1991

| Reciprocating | Screw | Vane | Power |
|---|---|---|---|
| 85 dBA | 76 dBA | 76 dBA | 0–7.5 kW |
| 89 dBA | 78 dBA | 76 dBA | 7.6–15 kW |
| 93 dBA | 80 dBA | 79 dBA | 15.1–22 kW |
| 97 dBA | 92 dBA | 90 dBA | 22.1–60 kW |

7.9 In noise-sensitive areas, an acoustic enclosure should be included in the purchase specification for all compressors. Such an enclosure should produce a reduction of at least 10 dBA in the free-field noise level at 1 m.

## Air intake

7.10 The position of an air intake can have a considerable effect on delivered air quality, particularly with respect to levels of carbon monoxide. The air intake for a compressor should be located to minimise contamination from internal combustion engine exhausts and the discharge from vacuum systems, AGSS and ventilation systems or other sources of contaminants. Air intakes should be ducted where necessary to avoid contamination; a minimum height of 5 m above ground level should ensure a reasonable quality of intake air. Where this cannot be achieved, additional filtration and/or air treatment may be necessary. If the siting of the compressor, regardless of the air intake location, is considered subject to a risk of aspirating toxic fumes and smoke as a result of a fire, an automatic shutdown system, linked to local smoke detectors, can be installed. If such a system is planned, it is essential that an automatic emergency supply manifold system is sited well away from the fire-risk area and is arranged to come on-line automatically in the event of plant shutdown.

7.11 Care is needed when extending compressor air intakes. Manufacturers' data should be consulted to ensure that intake flow, and hence compressor performance, are not adversely affected by excessive lengths of intake ducting. Choice of intake material is also important. Often, intakes are constructed from solvent-welded PVC. In a fire, toxic materials from the burning intake could be drawn into the air compressor and distributed throughout the system. In addition, there is a risk that inadequate solvent drying time before use of the intake will result in toxic solvent fumes being drawn into the system. Corrosion-resistant ducting (for example stainless-steel flue liner) is a suitable material.

7.12 Air-inlet filters should be fitted immediately upstream of the compressor. In exceptional circumstances, additional screens, filters and silencers may be required. The filters should comply with BS ISO 5011:2000 and be either dry medium filters or grade CA paper element filters.

## Compressor types

7.13 There are many different types of compressor currently available, the most common types being:

a. reciprocating piston compressors;

b. rotary vane compressors;

c. rotary screw compressors.

7.14 The compressors may be of any type, provided they are suitable for continuous running on load and for high frequency start/stop operation. When selecting compressors, the opportunity should be taken to maximise energy efficiency. If reciprocating compressors are used, they may be either of the single- or of the two-stage type, although for a 400 kPa system a single-stage compressor is usually satisfactory.

## Compressor lubrication

7.15 Compressors may be oil-lubricated, provided that suitable arrangements are made to ensure that the air quality specification given in Table 29 is fulfilled.

7.16 Rotary compressors are sealed and cooled by oil or water. Oil control is therefore essential and is usually provided as an integral part of the compressor. Reciprocating compressors may be oil-lubricated, carbon ring, PTFE ring or diaphragm-sealed type.

7.17 Oil-free compressors may be beneficial in reducing filtration requirements.

7.18 Water should not be used as a sealant because of risk of microbial contamination and potential problems with water treatment.

7.19 There is a danger that PTFE rings and lubricating oils could decompose at high temperatures to form toxic products. This may be countered by fitting a temperature sensor to the cylinder head or output of the compressor with suitable controls to cut off the power supply to the compressors if excessive temperatures are sensed. BS EN ISO 15001 specifies the requirements for selecting materials used in medical supply equipment.

7.20 On start-up, when oil is used as the sealant, moisture condensing at high pressure forms an emulsion with the oil. Once operating temperature is reached, water is readily separated. Because it is impossible to match the varying demand with plant capacity, it may be necessary to include oil heaters to avoid emulsification. If it is intended to omit oil heaters, manufacturers should be asked to confirm the suitability of the compressor for intermittent operation. Oil-lubricated compressors, however, are considered to be satisfactory.

7.21 Where oil-lubricated compressors are used, suitable means of separating oil from condensate is essential.

7.22 Once a compressor installation has been selected:

a. the plant should include at least two compressors, but additional compressors may be included provided that in all cases the total capacity will provide 100% of system design flow with one compressor not running;

b. the individual compressors should be arranged so that they will supply the system simultaneously if necessary;

c. the relative magnitude of the capital and running costs should be evaluated at the time of purchase. Too much emphasis has been placed on low capital cost at the expense of reliability and high power costs. The running costs should be calculated at realistic levels of usage;

d. the control system for the compressor plant should include an "hours-run" counter and should be constructed in accordance with the guidelines given below;

e. the efficiency of plant, expressed as the volume of air delivered to the pipeline distribution system (after losses in the drying system and filtration system) per kilowatt-hour (kWh), should be stated by the supplier of the system. The testing procedure should evaluate this efficiency by testing the power consumption over a suitable period of time at 100%, 10% and 0% of the system design flow. A minimum efficiency of 5 $m^3$/kWh at 100% and 10% is required. The power consumption at zero flow should be less than 1% of that at 100% design flow.

## After-coolers

7.23 After-coolers (and inter-coolers) usually form part of the compressor sub-assembly. After-coolers should be fitted to oil-lubricated medical air compressor systems. These will normally be air-cooled, and may need ducting with forced ventilation to ensure an adequate supply of cooling air.

## Receivers

7.24 Air receivers should comply with BS EN 286-1: 1998 for all vessels up to 10,000 bar litres, and should be supplied with test certificates. The minimum water capacity of the receivers should be 50% of the compressor output in 1 minute, stated in terms of free air delivered at normal working pressure. Receivers should also be fitted with an automatic drain. Electrically operated automatic drains have been found to be more reliable.

7.25 To facilitate the statutory inspection, there should be either two suitably valved air receivers or a by-pass arrangement (for use in manual operating mode only) in order to avoid interruption to the supply. Alternatively the tertiary supply manifold can be used.

7.26 For systems that have a design flow in excess of 500 L/min, two receivers should be provided with valve arrangements to permit isolation of one or the other for inspection purposes.

## Air treatment and filtration

### General

7.27 Contaminants can enter the compressed air system from three sources: the atmosphere, the compressor and the pipeline distribution system. Each potential source must be taken into account when specifying the type and location of air treatment equipment. Filtration equipment may include pre-filters, coalescing filters, adsorption equipment, carbon filters, particulate filters and any other additional filtration equipment necessary to ensure the quality of the product.

### Solid contaminants

7.28 Particles in the environment cover a wide range of sizes, but approximately 80% are less than 0.2 μm and are therefore not removed by the intake filter to the compressor.

7.29 Although particles smaller than 40 μm are unlikely to cause mechanical damage, a 5 μm intake filter is preferred to avoid blockage of internal air/oil separators.

7.30 Filters are specified in terms of performance tests – a sodium flame test, a DOP (dispersed oil particulate) test etc.

## Water

7.31 Water is always a contaminant in a compressed air system, regardless of the type and location of the compressor plant, since the air drawn into the compressor intake is never completely free of water vapour. The amount can vary from 2.5 $g/m^3$ to over 40 $g/m^3$ depending on the climatic conditions. The after-cooler and receiver remove some of this, but about 20 $g/m^3$ is likely to remain in the compressed air unless removed by dryers.

7.32 A water content not exceeding 67 vpm (volume parts per million – equivalent to dew-point –46°C at atmospheric pressure) is specified for medical air pipeline systems. Only desiccant dryers can usually achieve this. A variety of desiccant types are available. Activated alumina and silica gel are commonly employed. Molecular sieve desiccants employing zeolites can also be used, but on occasions it has been found that this material has produced air with an increased oxygen content, in the order of 24%. Refrigerant dryers can perform satisfactorily down to a pressure dew-point of +3°C (atmospheric dew-point –20°C) and are therefore not recommended as the sole form of drying.

## Oil

7.33 With oil-lubricated compressors, it is inevitable that the compressed air will contain oil. Even with oil-free compressors (non-lubricated), complete freedom from oil and oil vapour cannot be positively guaranteed, as hydrocarbon vapours may be drawn into the compressor. Oil levels in the air supply must be controlled to 0.1 $mg/m^3$ with means of monitoring on a routine basis.

7.34 Oil will exist in the system in three forms: bulk liquid, oil aerosol and oil vapour. Provided that the oil lubricant is appropriate and the after-cooler properly designed, the amount of oil present as vapour should be small and is unlikely to exceed 0.5 $mg/m^3$.

7.35 The amount of oil that is present as bulk liquid and aerosol is more difficult to predict. With modern, well-maintained oil-lubricated compressors, it is unlikely to exceed 5 $mg/m^3$ due to the high-efficiency oil/air separator.

7.36 Oil-contaminated compressor condensate is classified as a trade effluent by virtue of Chapter 14 of the Public Health (Drainage of Trade Premises) Act 1937. An oil condensate separator should therefore be installed.

7.37 Under Chapter 85 of the Water Resources Act 1991, it is illegal to make a discharge of trade effluent to "controlled waters" via a surface water drain without the consent of the Environment Agency.

7.38 Similarly, under the Water Industry Act 1991, local water companies enforce the limit of oil condensate discharged into the public foul sewer. Prior consent to discharge is mandatory.

7.39 Condensate from oil-free compressors may be discharged to drain.

7.40 Any condensate produced from the compressor/dryer system must be regarded as trade effluent and is therefore not suitable for discharge to any surface water system draining to any surface water sewer, water-course or soak away; this may not apply if a suitable separator is installed. Maximum oil content limits range from region to region, from 25 mg/L up to 500 mg/L; the local water company should be consulted.

## Dryer controls

7.41 The dryer control system should ensure that regeneration is operated in proportion to the compressed air usage. The effectiveness of the control system will become apparent when the efficiency of the compressor system is tested at 10% and 0% of the system design flow. Evidence of the reliability and performance of a dryer system should be sought from manufacturers, since these items are critical to the overall performance of the compressor system. The dryer control system should include a dew-point hygrometer and display with a minimum accuracy of ±3°C in a range from –20°C to –60°C atmospheric dew-point, with a set point of –46°C. It should be arranged that in the event of open circuit, a "plant emergency" alarm be initiated.

## Dust filters

7.42 There should be a dust filter downstream of the dryers to remove particles down to 1 μm, with a

DOP penetration of less than 0.03%, when tested in accordance with BS EN ISO 3549:2002.

7.43 Each dryer and filter assembly should be rated for continuous use at the system demand flow, with air at 100% relative humidity at 35°C.

### Activated carbon filter

7.44 Duplex activated carbon filters should be installed upstream of the final bacteria filter for odour removal.

### Bacteria filters

7.45 Duplex bacteria filters should be fitted upstream of the final pressure regulator with appropriate isolating valves. The filters should provide particle removal to 0.01 mg/m$^3$ and a DOP penetration of less than 0.0001%.

## Pressure control

7.46 The pressure control should maintain the nominal pipeline pressure within limits given in Chapter 4. Duplex line pressure regulators should be provided with suitable isolating valves. The regulators should be of the non-relieving type.

### Safety valves

7.47 Safety valves should be provided in accordance with the requirements given in (a)–(c) below. All safety valves should conform to BS EN ISO 4126-1:2004. A safety valve of the certified discharge capacity stated should be fitted in each of the following positions:

a. on the delivery pipe of each compressor and upstream of any isolating valve, non-return valve or after-cooler, capable of discharging the total throughput of the compressor;

b. on each air receiver and dryer tower, capable of discharging the sum of the throughput of all the compressors. It is not necessary to provide safety valves on the dryer columns where the system is already protected by a safety valve on the receiver and the downstream equipment, that is, if the dryer column is already sufficiently protected;

c. immediately downstream of each pressure regulator, capable of discharging the system demand flow.

7.48 All safety valves should be of the closed-bonnet type and connected to suitably sized pipework to allow safe discharge, not necessarily to the outside.

## Traps, valves and non-return valves

### Automatic drainage traps

7.49 Electrically- or mechanically-operated automatic drainage traps should be provided on the after-coolers, receiver, separators and coalescing filters. The discharge from these drainage traps should be piped to a suitable gully via an oil separator. Co-ordination with building work is required for this provision. Electrically-operated automatic drains have been found to be more reliable.

7.50 Drainage and tundishes are usually provided under the building contract. Separators should be provided under the air compressor contract. Provision of interceptor tanks may be made under either the building contract or the air compressor contract, as appropriate.

7.51 Non-return valves are required to prevent backflow of the air supply in certain situations. These valves should be located as follows:

a. between the compressor and the receiver, but downstream of any flexible connector;

b. downstream of the dust filter on the dryer;

c. upstream of the emergency cylinder reserve connection in the pipeline connecting the plant to the pipeline distribution system, to prevent back-feeding this plant;

d. upstream of any inlet point that may be used to feed the system in an emergency;

e. downstream of the emergency cylinder manifold regulators.

### Isolating valves

7.52 Isolating valves should be provided downstream of non-return valves and upstream of, for example, the connection of the emergency reserve manifold. Isolating valves should be provided in order to facilitate maintenance or replacement of plant items.

7.53 Manually-operated ball isolation valves should be located in the positions shown in Figures 23–26 to allow isolation of components such as receivers, dryers, automatic drains, pressure regulators and filters. There should also be a valve on the compressed air plant, downstream of the plant non-return valve and the connection of the cylinder manifold supply.

### Pressure indicators

7.54 Pressure indicators should comply with BS EN 837-1:1998 or have an equivalent performance if electronic indicators are used. Calibration should be in bar or kPa. All gauges should have a minimum scale length of 90 mm, and the working range should not exceed 65% of the full-scale range except on differential pressure gauges. Pressure indicators should be connected by means of gauge cocks.

7.55 Pressure indicators should be located:

a. on the plant control unit indicating receiver pressure;

b. on each receiver;

c. downstream of each pressure regulator;

d. on each dryer tower;

e. on the plant pipework, upstream of the plant isolating valve.

7.56 Differential pressure indicators should be located on:

a. each coalescing filter;

b. each dust filter;

c. each bacteria filter;

or any combination, as appropriate.

7.57 Except for pressure gauges, all control and measuring devices should be connected directly to the pipework via a minimum leak device (to allow removal for servicing) and not isolated by valves.

## Operating and indicating system

7.58 The operating and indicating system should perform the following functions:

a. overall plant control and indication;

b. individual compressor starting;

c. control of dryers;

d. plant status monitoring.

7.59 Provided that the individual compressor starters are housed in a separate compartment, these functions may be carried out by separate units or may be installed in a common panel and located on the plant or on the plantroom wall. Control panels containing pneumatic components should have vents to permit release of pressure in the event of component failure. All indicators should be appropriately identified and should have a design life of at least five years.

7.60 The operating system should be capable of automatically restarting after reinstatement of the power supply.

7.61 All components of the medical air supply system should be connected to the essential electrical supply. The control system should ensure that compressors restart in sequence to avoid overloading the power supply.

### Plant control unit

7.62 The plant control unit should have a separate power supply for each compressor, controlled by a separate sub-circuit.

7.63 The unit should allow either manual selection of duty/stand-by for each of the compressors or have an automatic sequence selection with a means for manual override. The unit should ensure that two or more compressors do not start simultaneously when power is applied.

7.64 A warning notice that complies with BS 5499-5: 2002 should be affixed which indicates the presence of low voltage.

### Plant control indication

7.65 There should be indicators for each compressor as follows:

a. green "mains supply on";

b. green "compressor called for", which indicates that the compressor motor is electrically energised;

c. an indicator of the pressure produced by the compressor.

### Compressor starter units

7.66 There should be individual starter units for each compressor which operate a single designated compressor. The starters should be provided with safety interlocks, as specified by the compressor manufacturers, which should inhibit plant operation until manually reset by means of a button. The starters should allow automatic restart after an interruption to the power supply. Each starter unit should contain the following:

a. an isolator interlocked with the covers;

b. either HRC (high rupturing capacity) fuses to BS 88 or suitable circuit breakers to BS EN 60947-2:2003 and/or BS EN 60898-1:2003;

c. an industrial grade ammeter to BS EN 60051-1:1999, IEC 60051-1:1997 (digital ammeters of similar accuracy to those compliant with BS EN 60051-1:1999, IEC 60051-1:1997 may be used);

d. a "total hours" counter if not included in the plant control unit;

e. a green "mains supply on" indicator if mounted separately from the plant control unit.

### Dryer control unit

7.67 The dryer control unit may be mounted on the dryers or may be located with the plant control unit. There should be separate power supplies for the duty and stand-by dryer assemblies taken from the same phase.

7.68 The dryer control unit should contain the following:

a. a duty dryer selector switch;

b. a service function – to enable selection of continuous/normal running;

c. individually fused, separate cycling systems for each dryer;

d. a system to control regeneration of the dryers in relation to pipeline demand;

e. a hygrometer and display with a minimum accuracy of ±3°C in a range from –20°C to –60°C (set to –46°C atmospheric dew-point) and a pressure sensor;

f. an automatic changeover to the stand-by dryer system in the event of failure of the duty unit by either dryness or pressure. This requires:

   (i) electrical and pneumatic isolation of the duty sub-assembly so that it is taken off-stream;

   (ii) electrical and pneumatic energisation of the stand-by sub-assembly so that it is brought on-stream;

   (iii) activation of the appropriate fault indicator and associated volt-free contacts;

   (iv) the sub-assembly to remain in this mode of operation until the fault has been rectified;

> **Note**
>
> In the event of power supply failure, all drain and vent valves should fail "closed", and all inlet and outlet valves should fail "open".

g. green function indicators for each dryer sub-assembly to indicate:

   (i) dryer 1 selected;

   (ii) dryer 2 selected;

   (iii) selected dryer – "normal";

   (iv) selected dryer – "failed" (this fault indicator should remain until manually reset by means of a reset button);

h. a fail-safe system which on failure of the power supply causes the following:

   (i) closure of the exhaust and purge valves;

   (ii) opening of the inlet and outlet valves.

### Plant status monitoring

7.69 A monitoring system should be provided to detect the following faults in the air compressor system:

a. plant faults (for each compressor):

   (i) control circuit failed;

   (ii) motor tripped;

   (iii) after-cooler temperature high;

   (iv) compressor temperature high;

   (v) compressor failed to go on load;

   (vi) activation of other safety devices supplied by the manufacturers;

b. plant faults (for each dryer unit):

   (i) dryer failure;

   (ii) pressure fault;

c. plant emergency:

   (i) receiver pressure 0.5 bar below the stand-by cut-in pressure;

   (ii) receiver pressure 0.5 bar above cut-out pressure;

   (iii) dryness above –46°C at atmospheric pressure;

d. pressure fault (cylinder reserve):

(i) pressure in duty bank below 50% (of normal cylinder pressure);

e. pressure fault (pipeline):

(i) low pipeline pressure;

(ii) high pipeline pressure.

### Plant status indicator unit

7.70 In addition to the plant control indication, there should be a plant status indicator panel that may be mounted on the plantroom wall or adjacent to either the compressor starter unit or the plant control unit. It should have a warning notice that complies with BS 5499-5:2002 to indicate the presence of low voltage.

7.71 There should be indicators for each compressor to show the following conditions:

a. green "mains supply on";

b. yellow "control circuit failed";

c. yellow "overload tripped";

d. yellow "after-cooler temperature high";

e. yellow "compressor temperature high";

f. yellow for each individual safety device provided by the manufacturers;

g. yellow "compressor failure".

7.72 There should be indicators for each dryer system to show the following:

a. green "mains supply on";

b. yellow "dryness fault";

c. yellow "pressure fault".

### Alarm signal status unit

7.73 An alarm signal status unit should be provided as part of the control system. It should display the following conditions:

a. green "normal" (normal);

b. yellow "plant fault" conditions ((b)–(g) in paragraph 7.71);

c. yellow "plant emergency" (low reservoir pressure/high moisture: that is, condition (b) in paragraph 7.71);

d. yellow "reserve low" (emergency/reserve banks low (<50%));

e. red "pipeline pressure fault" (pressure fault).

7.74 Conditions (b) to (e) should be transmitted to the central alarm system. Where relays are used, they should be normally energised relays that de-energise under fault conditions, with contacts having a minimum rating of 50 V dc and 50 mA.

7.75 Volt-free, normally closed contacts rated at 50 V dc and 50 mA should be provided for transmission of conditions (b) to (e) to the alarm system.

7.76 The panel can be incorporated into the plant indicator unit or be a separate unit within the plantroom. If mounted separately, the cabling should be monitored for open/short circuit. In the event of such a cabling fault, a red "system fault" lamp should be illuminated on the alarm signal status unit together with the appropriate alarm condition.

7.77 The alarm signal status unit should be supplied from all individual plant control units or from a separate common supply.

### Plant management

7.78 Connections should be provided which allow monitoring of plant alarm conditions (b) to (e) and pump running for each "compressor". These connections should be volt-free contacts normally closed for each condition having a minimum rating of 50 V dc and 50 mA. The building management system should not be used to control the plant.

## Synthetic air

7.79 This section provides technical details of the process and systems required to generate medical air from mixing gaseous oxygen and nitrogen, derived from cryogenic supplies.

7.80 For the purposes of the Medicines Act 1968, it is considered that the synthetic air is manufactured on-site, for use on that site only, in exactly the same way as for medical air derived from compressor plant. The production of synthetic air implies a manufacturing process, and as such, the process should be subjected to the same safety requirements of any pharmaceutical process. This should include, for example, a HAZOP (HAZard and OPerability) analysis and other safety analyses that may be necessary.

7.81 Synthetic air is generated by mixing gaseous oxygen and nitrogen in a blender or mixing panel at pre-set pressures to ensure that the resultant mixture is always correct. Continuous on-line monitoring of oxygen concentration is provided to check the mixture; the system shuts down automatically if the oxygen concentration varies from the specified value.

7.82 If one mixing system shuts down, the pipeline is supplied from the secondary mixing system to ensure continuity of supply.

7.83 The feasibility study should provide more information on the details of the monitoring and alarm systems required, as well as operational information.

7.84 The VIE system supplying the medical oxygen may be used to supply the synthetic air system, depending on the system demands.

7.85 Nitrogen supplied to the synthetic air system may also be used to provide the power source for surgical tools instead of surgical air at 700 kPa.

7.86 An electrical power supply is required in order, for example to operate solenoid valves and monitoring instrumentation. Therefore the system should be connected to the essential power supply and via an uninterruptible power supply (UPS) with at least four hours' capacity; this should ensure continuity of supply in the event of power failure.

## System description

7.87 The gaseous oxygen and nitrogen are derived from bulk liquid supplies contained in a VIE – as described in the "Liquid oxygen systems" section of Chapter 6.

7.88 The oxygen for synthetic air may be taken from the VIE supplying the medical oxygen system or it may be from a dedicated VIE. It would normally be more cost-effective for the oxygen to be taken from the main VIE, although this would obviously depend on the existing VIE capacity, the demand, space constraints etc. The feasibility study should provide more detailed information on whether it is likely to be more cost-effective to provide a totally separate VIE system or to use the existing medical oxygen VIE.

7.89 For both the oxygen and nitrogen it is necessary to have a secondary supply system to ensure continuity of supply; the system demands are such that this should be derived from a second – normally smaller – VIE.

7.90 This secondary oxygen supply can also serve the hospital's medical oxygen system.

7.91 Since four VIEs will be required, the space requirements will need special consideration when planning the installation of a synthetic air system.

7.92 The system comprises:

a. storage vessels – one main vessel and one secondary supply vessel for both oxygen and nitrogen;

b. vaporisers for both oxygen and nitrogen;

c. medical oxygen flow control – where used to supply medical oxygen systems;

d. surgical nitrogen flow control – where required;

e. a control panel for the nitrogen and oxygen supplies to the mixing panels;

f. duplicate air mixing panels;

g. buffer vessels – each mixer has a buffer vessel to smooth fluctuations in demand;

h. a warning and alarm system;

j. duplicate oxygen analysers on each mixer.

7.93 The system is shown in Figure 27.

## Storage vessels

### Vessel summary

7.94 The following vessels are required:

a. one main oxygen vessel;

b. one secondary oxygen vessel with at least 24 hours' capacity;

c. one main nitrogen vessel;

d. one secondary nitrogen vessel with at least 24 hours' capacity.

### Vessel operating pressure

7.95 The following operating pressures are required:

a. main vessels: 12.5 bar;

b. back-up vessels: 12.5–14 bar.

### Main vessel capacity

7.96 The main vessel should normally be sized on the basis of two weeks' supply. This should be calculated as 14 x the average daily usage. This should provide adequate storage and a cost-effective vessel-filling regime. The gas supplier should, however, be consulted as there may be other factors, such as geographical location, space etc, which need

# Figure 27 Synthetic air plant

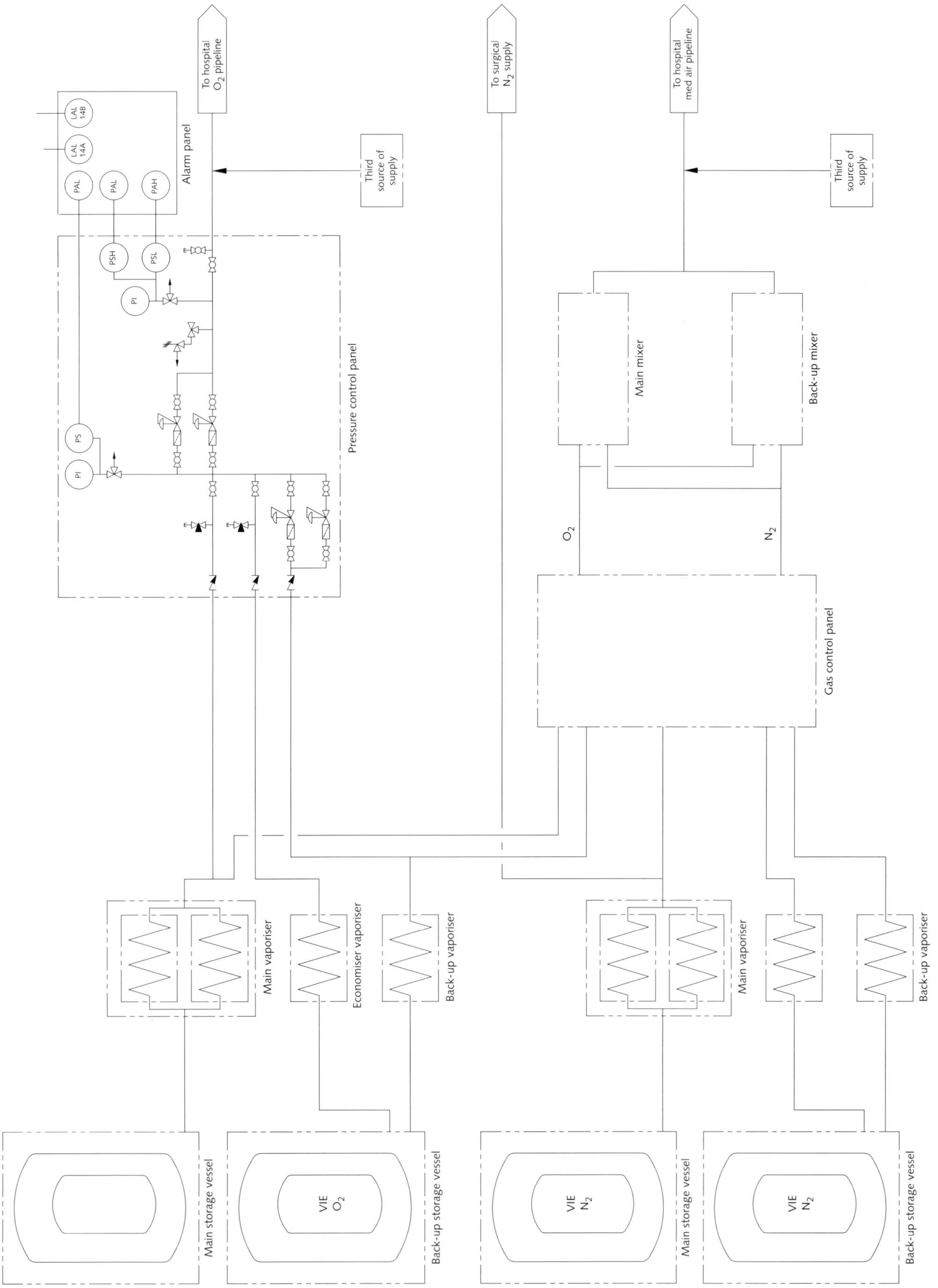

to be taken into account when sizing the main vessels.

### Back-up vessel capacity

7.97 The stand-by vessel should have 24 hours' capacity at any time; that is, it should be sized on the basis of twice the average daily usage. This will ensure that there is always 24 hours' supply available.

7.98 In addition to the normal instrumentation as set out in the "Liquid oxygen systems" section of Chapter 6, the vessels should be fitted with a telemetry system to continuously monitor the vessel contents.

7.99 This information should be transmitted direct to the gas supplier and also the hospital. The exact details of how much information, and where it should be received, will depend on each hospital site.

7.100 The main vessel low level alarm is activated at 25% full; the back-up low level alarm is activated at 50% full.

7.101 The safety relief valves and bursting discs should be sized in accordance with BCGA CP19.

7.102 The liquid from the vessels should be supplied to the process at a nominal pressure of 12.5 bar.

## Vaporisation

7.103 The main and stand-by vessels should have dedicated vaporisers designed for continuous capacity and 24-hour capacity respectively at 1.5 x the required flows to ensure that the vaporisers are not overdrawn.

7.104 This may be achieved in each case by either a single set of vaporisers or by vaporisers operated on timed or manual changeover.

7.105 It is preferable for the vaporisers to operate on a timed changeover as this avoids the need for hospital staff to manually operate the changeover valves.

7.106 The timed changeover will require a 110 V or 240 V supply; this should be on the emergency supply and a UPS should also be provided, with at least 4 hours' capacity.

7.107 Each vaporiser or set of vaporisers must have a safety relief valve.

## Medical oxygen flow control

7.108 A control panel (similar in principle to a C11 panel) should be provided – the only difference is that the secondary supply is taken from a low-pressure liquid source.

## Surgical nitrogen flow control

7.109 A control panel to regulate the gaseous nitrogen to between 7.5 and 9.5 bar, depending on the system design, should be provided.

7.110 The pipeline distribution system should be designed in exactly the same way as for surgical air 700 kPa systems, as described in Chapter 8.

## Control panel for the nitrogen and oxygen supplies to the mixing panels

7.111 The control panel should be sized to provide pressure-regulated flows as appropriate for the mixing system; this would typically be up to 200 $Nm^3/hr$ (normal cubic metres per hour).

7.112 The stand-by supply regulation cuts in when the main line pressure falls to 11 bar; there is no regulation on the main supply line.

7.113 A non-return valve should be installed in both the nitrogen and oxygen supply lines within the mixer to prevent cross-contamination.

7.114 A non-return valve should also be installed on both the main oxygen supply and the stand-by oxygen supply to the mixer to prevent the medical oxygen line becoming contaminated with nitrogen.

## Air mixing panels

7.115 A range of sizes of mixing panels is available with, typically, nominal capacities of 50, 100 and 200 $Nm^3/hr$.

7.116 A regulated supply of nitrogen and oxygen is blended in a mixing valve. The differential pressure at the inlet to the mixing panel is critical and should not exceed 0.5 bar. A pressure-switch-operated solenoid valve opens and shuts on a 0.5 bar differential.

7.117 The main mixer solenoid valve opens when the line pressure falls to 4.2 bar; the stand-by mixer solenoid valve will open if the line pressure continues to fall to 4.0 bar.

7.118 Two independent paramagnetic oxygen analysers are provided on each mixer to give continuous on-line measurements.

7.119 If the oxygen concentration falls outside 20–22% as measured by either analyser, the mixer solenoid valve is held closed and the mixer is shut down. In addition, a signal is relayed downstream to close the solenoid valve on the buffer vessel associated with that mixer.

## Buffer vessels

7.120 Each mixer has associated with it a buffer vessel to smooth fluctuations in demand.

7.121 In the event that the oxygen concentration differs from the specification (that is, 20–22%), the solenoid valve downstream of the buffer vessel will also close, preventing air from the buffer vessel from entering the distribution system.

7.122 The buffer vessel, together with appropriate means of safety relief, should be sized to match each mixing panel to provide stable operation.

## Alarm signal status unit

7.123 The same alarm conditions for liquid oxygen should also be transmitted and displayed for the liquid nitrogen system. The following conditions should be displayed for the mixing panels:

a. green "normal" (normal);

b. yellow "plant fault" (low gas pressure to any mixer);

c. yellow "plant emergency" (analysis out of specification on any mixer);

d. yellow "reserve low" (operating on final mixing panel/buffer vessel only);

e. red "pressure fault" (pressure fault).

7.124 Conditions (b) to (e) should be transmitted to the central alarm system. Where relays are used, they should be normally energised relays that de-energise under fault conditions, with contacts having a minimum rating of 50 V dc and 50 mA.

7.125 Volt-free, normally closed contacts rated at 50 V dc and 50 mA should be provided for transmission of conditions (b) to (e) to the alarm system.

7.126 The panel can be incorporated into the mixing panel control unit or be a separate unit within the plantroom. If mounted separately, the cabling should be monitored for open/short circuit. If such a cabling fault occurs, a red "system fault" lamp should be illuminated on the alarm signal status unit together with the appropriate alarm condition.

## Emergency supply provision

7.127 A risk assessment should be carried out to establish the vulnerability of the main supply system of both oxygen and nitrogen. Further information is given in Chapter 2 on sources of supply and in Chapter 6.

## Additional use of medical air systems

7.128 It is possible to use medical/surgical air as a power source for pendant control and braking systems.

7.129 These additions must not compromise either the medical air system or operation of connected equipment. They must be connected via a non-return valve and flow-limiting device, and be capable of isolation by means of an AVSU labelled to identify the equipment controlled.

7.130 Medical air systems must not be used to provide air for sterilizer chamber or door-seal use.

# 8 Surgical air systems

## General

8.1 Surgical air at 700 kPa is only used as the power source for surgical tools. These tools typically require high flows – up to 350 L/min – at 700 kPa at the point of use. Where nitrogen is available on site, it may be used as an alternative source of supply.

8.2 Supply systems for surgical compressed air may be a cylinder manifold system, a dedicated 700 kPa compressor system or a compressor system capable of supplying both the 700 kPa and the 400 kPa supplies. In practice, the decision about which compressor system to install needs careful consideration because of the flow rates required and total usage (see Chapter 7).

8.3 A compressor system will be required for large operating department complexes specialising in orthopaedic and/or neurosurgery that require the use of pneumatically-powered surgical tools. An automatic reserve manifold located in separate accommodation should be provided. A typical system is shown in Figure 28.

8.4 It is possible to use nitrogen instead of air as the power source for surgical tools. This may be derived from either a liquid source or cylinders. In either case, the terminal units must be different from the existing medical air 700 kPa terminal units. A NIST connector is already specified for nitrogen and should be used.

8.5 The pressure control equipment should comprise duplex regulating valves with upstream and downstream isolating valves, pressure gauges and pressure relief valves.

8.6 Whatever supply system is installed, the overall system should be designed to provide a minimum of 700 kPa at the front of each terminal unit at a flow of 350 L/min.

### Note

Systems designed to meet requirements of earlier editions of Health Technical Memorandum 2022 may not provide 350 L/min at 700 kPa. Information on upgrading surgical air systems in given in Appendix J.

8.7 The maximum pressure at the terminal unit under "static flow" conditions should not exceed 900 kPa.

8.8 Cylinders of medical air or nitrogen stored locally should always be available for use in an emergency.

8.9 Vessels should be selected as follows:

| Design flow (L/min) | Vessel size | Compressor output (L/min) |
|---|---|---|
| >500 | 1 × 200% design flow | 0.33 × design flow |
| 500–2000 | 2 × 66.6% design flow | 0.66 × design flow |
| 2000–3500 | 2 × 200% design flow | 0.66 × design flow |
| 3500–7000 | 3 × 33.3% design flow | 0.5 × design flow |

## Extension of surgical air systems into dental departments

8.10 Some surgical air systems have been extended into dental departments; such an extension offers obvious economic and air quality advantages in comparison with separate provision. When such extensions are made, a non-return valve or back-feed protection device, with upstream and downstream isolating valves, should be installed in the supply line to the dental department. Before extending a surgical air system into a dental department, the following must be taken into account:

a. the extra demand on the existing system must not compromise patient safety or operation of either the existing system or its extension. In particular, the ability of an existing emergency supply system to cope with potentially very high demands must be carefully assessed;

b. the Authorised Person (MGPS) with responsibility for the existing surgical air system

## Figure 28 Typical simplex surgical air plant and automatic emergency reserve manifold

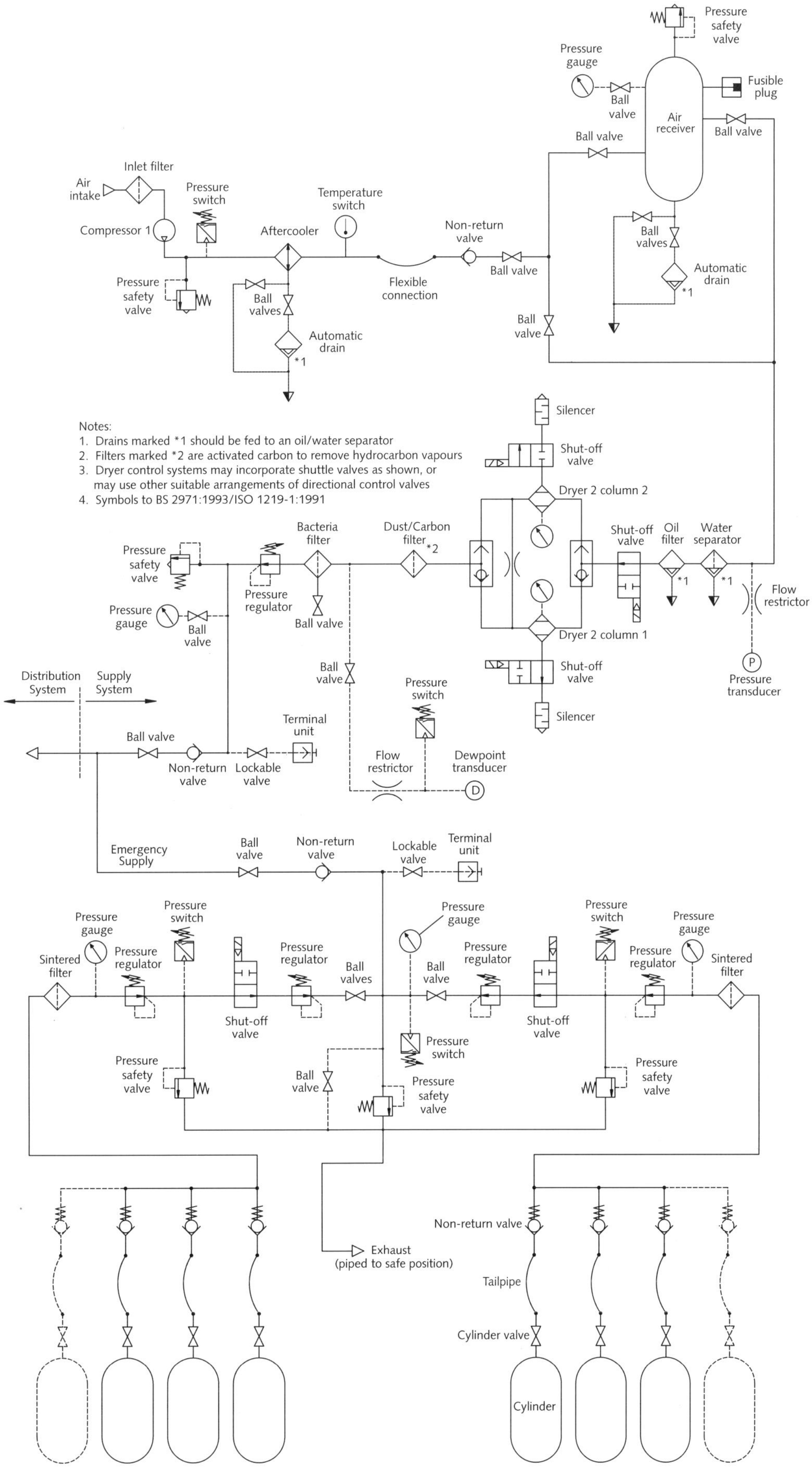

will automatically assume responsibility for the whole of the dental compressed air and vacuum system. Both the Authorised Person (MGPS) and Quality Controller (MGPS) must appreciate that extending a surgical air system into a dental unit for dental instrument use will introduce "non-standard" pipework terminations, for example crimped or compression-fitted connectors, in addition to non-degreased components. Failure of these "non-standard" components could lead to a serious depressurisation of the existing surgical air system and, if provided from the same source, the associated medical air system. If the system is further extended into a dental laboratory, surgical air could be used to support the operation of such devices as natural gas/air burners. Cross-connection of these systems is unlikely, but the risk must be assessed;

c. if the surgical air is derived from a plant that supplies medical air, the medical air supply should have a separate manifold reserve supply when space and system design makes this practicable.

# 9 Medical vacuum systems

## General

9.1 The medical vacuum pipeline system provides immediate and reliable suction for medical needs, particularly in surgical accommodation.

9.2 The medical vacuum pipeline system consists of the vacuum supply system, the distribution pipework and terminal units. The performance of the pipeline system is dependent on the correct specification and installation of its component parts. This chapter describes the requirements of the vacuum supply system.

9.3 The medical vacuum pipeline system should be designed to maintain a vacuum of at least 300 mm Hg (40 kPa) at each terminal unit during the system design flow tests.

9.4 To ensure continuity of supply, the vacuum plant should be connected to the essential electrical power supply.

9.5 The capacity of the vacuum supply system should be appropriate to the estimated demand.

9.6 With the exception of the vacuum discharge to atmosphere, the pipeline distribution system for vacuum has traditionally been constructed of copper. PVC pipework can be considered where cost-effective. Pressure testing of PVC and copper pipework should be carried out at 100 kPa.

9.7 The major components of a medical vacuum system and their layout are shown in Figure 29. A suitable operating and indicating system with alarms is also required. The location of the components should allow adequate space for access for maintenance. Packaged supply systems are available from manufacturers that should be specified to meet the requirements given in this memorandum.

9.8 The plant should consist of at least three identical pumps, a vacuum reservoir with by-pass facilities, duplex bacteria filters with drainage traps, appropriate non-return valves, isolating valves, gauges and pressure switches, an operating and indicating system, an exhaust system and a flow test connection. For capacities in excess of 500 L/min, two vessels that can be independently isolated should be installed.

### Note

The third means of supply for a vacuum installation will comprise portable suction units.

## Siting

9.9 The plant should have all-round access for maintenance purposes, and allowance should be made for changing major components.

9.10 The siting of the plant should allow for adequate flows of air to cool the pumps. The manufacturers should be consulted over the range of operating temperatures for which the supply system is designed. In extreme cases, refrigerator cooling may be required.

## Pump noise

9.11 The noise level produced by the pumps will increase with the capacity of the supply system. For larger systems this can result in an unacceptable noise level at the pump. The maximum free-field noise level at 1 m from the unsilenced pump should not exceed the following values for individual pumps:

| Power (kW) | Noise level (dBA) |
|---|---|
| 5 | 75 |
| 5.1–15 | 82 |
| 15 | 89 |

9.12 A suitable acoustic enclosure may be required in the purchase specification for all pumps with a free-field noise level at 1 m of 80 dBA or over. An enclosure should produce a reduction of at least 10 dBA in the free-field noise level at 1 m.

## Figure 29 Triplex vacuum system

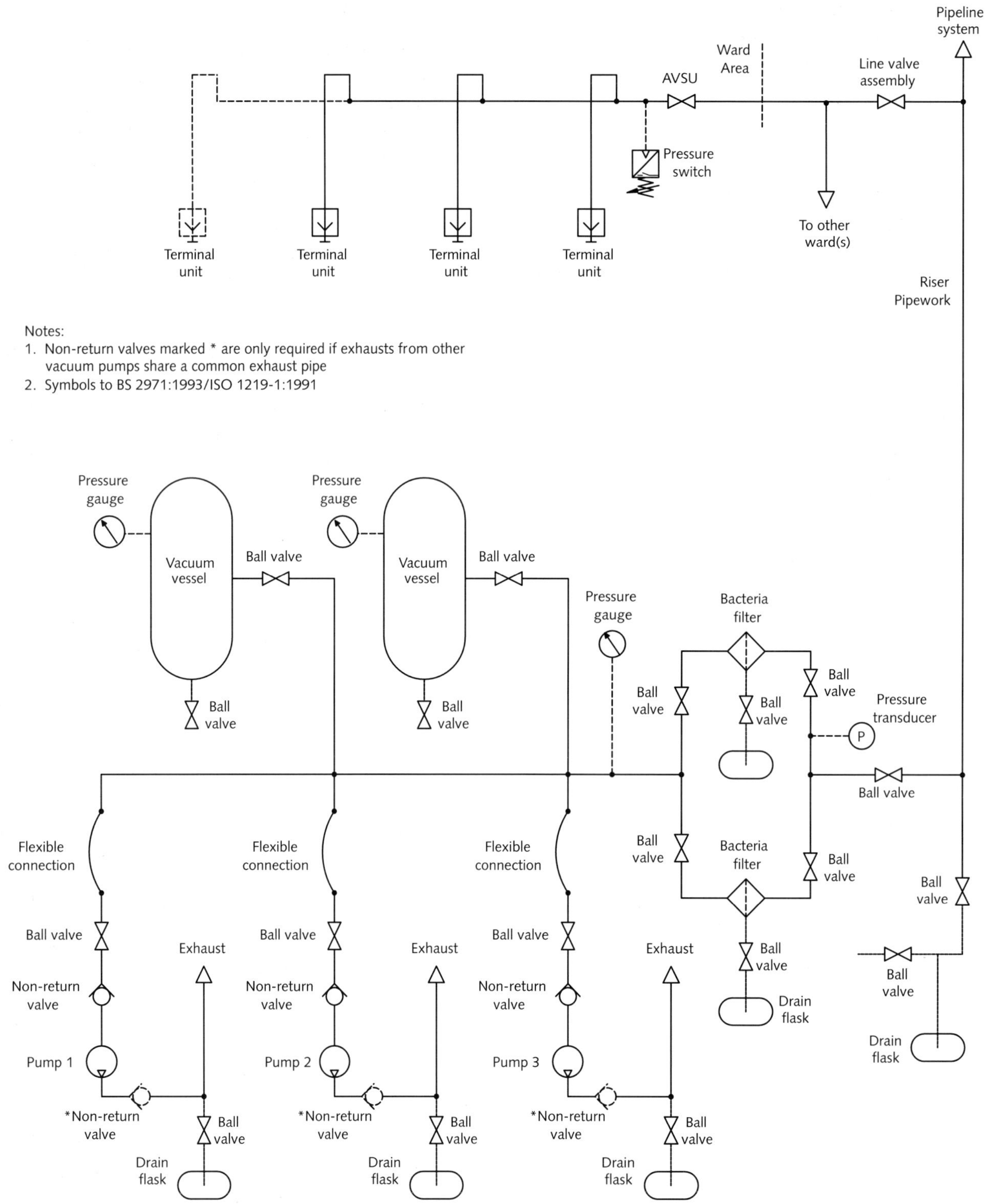


## Vacuum plant exhaust

9.13 The position of the termination point should be carefully chosen to be clear of windows, ventilation intakes and the intake of air compressors and other equipment, since for oil-lubricated pumps the vacuum exhaust is likely to be polluted with oil fumes.

9.14 Noise from the exhaust should be considered and a silencer fitted if necessary.

9.15 The construction should conform to the following criteria:

a. the exhaust should be sized to give a back pressure at system design flow which is matched to the pump performance;

b. the termination point should be turned down and provided with protection to reduce the effect of wind pressure and prevent the ingress of rain, snow, insects or animals;

c. weatherproof notices should be fixed at the discharge point(s) with the legend "medical vacuum discharge point – do not obstruct";

d. the exhaust pipe should be provided with a drainage valve at its lowest point;

e. a silencer should be fitted in the exhaust pipe from each pump. This may be integral with the pump unit.

## Efficiency

9.16 The pump should be capable of producing a higher vacuum than that required in the pipeline, so that the resistance of the bacteria filter and back pressure in the exhaust system can be overcome.

9.17 The capacity of the vacuum pump should be specified in terms of the free air aspirated (FAA) in L/min when the pump is operating at a vacuum of 475 mm Hg (63 kPa) and at 450 mm Hg (60 kPa) at the plant pipeline connection.

## Vacuum pumps

9.18 Any type of pump apart from water-sealed pumps can be used.

9.19 Pumps should normally be oil-lubricated. Vapours from the lubricating oil are unlikely to be a significant component of the exhaust gases if correctly maintained. "Dry running" rotary vane pumps are available at increased capital cost and with lower efficiency than oil-lubricated pumps of comparable performance.

9.20 At least three pumps should be provided. The actual number is at the discretion of the plant manufacturer to ensure optimum cost benefit of the system. All pumps should be designed for high frequency stop/start or continuous operation. The opportunity to maximise energy conservation should be taken into consideration.

9.21 All systems should comprise pumps and motors of identical type that are suitable for continuous running and stop/start operation.

9.22 Pump motors should comply with the National Health Service Model Engineering Specification C51 – 'Electrical requirements for specified equipment' with the addition of Class F insulation and Class B temperature rise.

9.23 A vacuum reservoir should be provided so that the duty pump does not run continuously for low loads. The reservoir should be manufactured in accordance with BS EN 286-1:1998, with test certificates provided to the user. The minimum test pressure should be 4 bar.

9.24 The water capacity of the reservoir should be equal to the plant design flow at 450 mm Hg (60 kPa) in terms of free air aspirated in one minute with the pump operating at 475 mm Hg (60 kPa).

9.25 Provision should be made for draining the reservoir under vacuum conditions. By-pass facilities should be provided so that the reservoir can be drained and inspected without interruption to the vacuum supply. The reservoir should be fitted with suitable lifting lugs and feet.

9.26 If multiple reservoirs are provided, they should be arranged in parallel.

9.27 The bacteria filters and drainage trap should comprise two identical sub-assemblies with manually-operated isolating valves, arranged to allow either sub-assembly to be on stream. Each sub-assembly should contain a bacteria filter rated at the plant capacity.

9.28 The bacteria filter should be marked with the legend "bio-hazard", together with a description of a safe procedure for changing and disposing of the filters and emptying the drainage trap.

9.29 The bacteria filters should have a filter efficiency, when tested by the sodium flame test in accordance with BS 3928:1969, of greater than 99.995% at the system design flow.

9.30 The pressure drop across a clean filter at the system design flow should not exceed 25 mm Hg (3 kPa) at a vacuum of 475 mm Hg (63 kPa).

9.31 The drainage trap may be integral with the bacteria filter and should be fitted with a transparent bowl to collect liquid. The bowl should be suitable for steam sterilization at 134°C.

9.32 Although there is no firm evidence that has demonstrated the need for bacteria filters, it is recommended that such devices are included as precautionary measures.

## Pressure control

9.33 The cut-in setting for the vacuum pumps should be adjusted to allow for the pressure drop across the pipeline distribution system and the bacteria filters. The cut-in may be expected at about 500 mm Hg (67 kPa).

9.34 The cut-out setting should be at an appropriate point on the performance curve of the pump, which minimises stop/start operation but is at a vacuum which is economically attained by the pump. This cut-out setting may be expected at about 650 mm Hg (87 kPa).

## Valves

9.35 Non-return valves should be fitted, when necessary, at the inlet and outlet of each pump to prevent backflow when a common discharge pipe is used. (Some vacuum pumps include integral non-return valves.)

9.36 Manually operated valves should be arranged in the positions shown in Figure 29 to allow isolation of components such as pumps, reservoirs, by-pass pipework, drainage taps and bacteria filters.

## Pressure regulation of vacuum system

9.37 A vacuum of 300 mm Hg is required at the connection point of each terminal unit with a flow of 40 L/min whilst the system is operating at system design flow.

9.38 This performance is tested by the procedures carried out in accordance with Chapter 15.

9.39 A pressure drop of 13 kPa (100 mm Hg) is allowed across the terminal unit at a flow of 40 L/min (BS 5682:1998). The minimum pressure at the front of the most distal terminal unit should be 40 kPa (300 mm Hg) at a flow of 40 L/min. The minimum pressure (dynamic) at the plant should be 60 kPa (450 mm Hg).

### Note

Precautions for changing filters are included in Part B.

## Vacuum indicators

9.40 Vacuum indicators should comply with BS EN 837-1:1998 or have an equivalent performance if electronic indicators are used. Calibration should be 0–760 mm Hg (0–101 kPa). All gauges should be a minimum scale length of 90 mm.

9.41 Vacuum indicators should be located on:

a. the plant control unit indicating the vacuum in the pipeline (that is, on the pipeline side of the bacteria filter);

b. each reservoir.

9.42 A differential vacuum indicator (to indicate filter blockage rather than quantitative pressure drop) should be located across the bacteria filter and have a service isolation valve.

## Electrical supply

9.43 The electrical supply to the medical vacuum plant should be connected to the essential electrical supply. A time-delay system should be provided to avoid overloading the power supply on changeover.

## Pump operating and indicating system

### General description

9.44 The operating and indicating system should perform the following functions:

a. overall plant control and indication;

b. individual pump starting;

c. plant status monitoring and indication;

d. alarm signal status unit.

9.45 Provided that the individual pump starters are housed in a separate compartment, the operating and indicating system may be housed in separate units or may be installed in a common panel and located on the plant or on the plantroom wall.

9.46 Pneumatic components should have ventilation. All functions should be appropriately identified. Indicators should have a design life of at least five years. The operating system should be capable of automatically restarting after reinstatement of the power supply.

9.47 The vacuum supply system should be connected to the stand-by electrical supply. The control system should ensure that pumps restart in sequence to avoid overloading the power supply.

### Plant control unit

9.48 The control unit should have a separate power supply for each pump controlled by a separate sub-circuit. It should be manufactured and installed in accordance with IEE regulations, and the design should be such that no single component failure in the control unit will result in loss of plant output.

9.49 The unit should allow either manual selection of duty/stand-by for each of the pumps or have an automatic sequence selection with a means for manual override. The control unit should ensure that two or more pumps do not start simultaneously when power is applied.

9.50 A warning notice which complies with BS 5499-1: 2002 should be affixed which indicates the presence of low voltage.

9.51 For testing purposes, each pump should have a selector switch which when turned to the "on" position allows the pump to run continuously.

### Plant control indication

9.52 There should be indicators for each pump as follows:

a. green "mains supply on";

b. green "pump operating", which indicates that the pump motor is electrically energised;

c. green "pump operating", which indicates that the pump is drawing vacuum;

d. an indicator of the vacuum produced in the pipeline.

### Pump starter units

9.53 There should be individual starter units, each one operating a single designated pump. The starters should be provided with safety interlocks as specified by the pump manufacturers, which should inhibit plant operation until manually reset by means of a button. The starters should allow automatic restart after an interruption to the power supply. Each starter unit should contain the following:

a. an isolator interlocked with the covers;

b. either HRC fuses to BS 88 or suitable circuit breakers to BS EN 60947-2:2003 and/or BS EN 60898-1:2003;

c. starter;

d. an industrial grade ammeter to BS EN 60051-1: 1999, IEC 60051-1:1997 or an electronic digital instrument of comparable, or higher, standard;

e. a total hours counter, if not included in the plant control unit;

f. a green "mains supply on" indicator, if mounted separately from the plant control unit.

### Plant status monitoring

9.54 A monitoring system must be provided to detect the following faults in the vacuum supply system:

a. plant faults for each pump:

(i) control circuit failed;

(ii) motor tripped;

(iii) pump failed to go on load;

(iv) activation of other safety devices supplied by the manufacturers;

b. plant emergency – receiver vacuum has fallen, for example, by 50 mm Hg below the cut-in setting for the pump;

c. pressure fault (pipeline) – pipeline vacuum less than 360 mm Hg.

### Plant status indicator unit

9.55 In addition to the plant control indication, there should be a plant status indicator panel that may be mounted on the plantroom wall or adjacent to either the pump starter unit or the plant control unit. It should have a warning notice that complies with BS 5499-1:2002 to indicate the presence of low voltage.

9.56 There should be indicators for each pump to show the following conditions:

a. green "mains supply on";

b. yellow "control circuit failed";

c. yellow "motor tripped";

d. yellow for each individual safety device provided by the manufacturers;

e. yellow "pump failure".

### Alarm signal status unit

9.57 The following indication of plant conditions should be provided:

a. green "normal" (indicator normal);

b. yellow "plant fault" conditions (b)–(d); see paragraph 9.56;

c. yellow "plant emergency" condition (e); see paragraph 9.56;

d. red "pipeline pressure fault" (pressure fault).

9.58 Conditions (b) to (d) should be transmitted to the central alarm system. Where relays are used, they should be normally energised relays, which de-energise under fault conditions, with contacts having a minimum rating of 50 V dc and 50 mA.

9.59 Volt-free, normally closed contacts rated at 50 V dc and 50 mA should be provided for transmission of conditions (b) to (e) to the alarm system.

9.60 The panel can be incorporated into the plant status indicator unit or be a separate unit within a plantroom. If mounted separately, the cabling should be monitored for open/short circuit. In the event of such a cabling fault, a red "system fault" lamp should be illuminated on the alarm system status unit together with the appropriate alarm condition.

## Plant management

9.61 Connections should be provided which allow monitoring (but not control) of plant alarm conditions (b) to (e) and pump running for each vacuum pump. These connections should be volt-free contacts normally closed for each condition having a minimum rating of 50 V dc and 50 mA.

9.62 Plant should be operated in accordance with the manufacturer's instructions and be covered by a sound, effective planned preventative maintenance (PPM) policy.

# 10 Anaesthetic gas scavenging disposal systems

## Terminology

**10.1** An active system, as specified in either BS 6834: 1987 or BS EN 737-2:1998, is one in which a high air flow generated by an electrically driven pump is used to exhaust air through the system's fixed pipework. This in turn entrains waste gases from the patient, or patient ventilator, via a transfer hose and receiving system.

**10.2** The transfer and receiving system form part of the anaesthetic/breathing system.

**10.3** The receiving system is designed to match the variable flow in the breathing system to the constant flow of the disposal system and ensure that very low induced flows are imposed (0.5 L/min in the case of BS 6834:1987 and 0.05 L/min in the case of BS EN 737-2:1998 systems).

**10.4** As a passive system essentially comprises a pipe through a hole in a wall through which waste gases are driven by the patient or ventilator expiratory effort, there is no pump involved in such a system. In the UK, only systems complying with the BS or EN Standards above are considered appropriate for scavenging waste anaesthetic gases from accommodation in which general anaesthesia is taking place.

**10.5** Active scavenging for dental installations is an entirely different concept. An active system is one in which there is a flow generated through the patient's nasal mask and this carries away the waste gases exhaled by the patient. This flow is in the order of 45 L/min and is achieved by connection of the mask (via a suitable flow-limiting adaptor) to either a dental vacuum system or directly to an active scavenging system (BS/EN) wall terminal unit.

## General

**10.6** Anaesthetic gases are considered to be substances hazardous to health for the purposes of the Control of Substances Hazardous to Health Regulations 2002 (COSHH), except where they are administered to a patient in the course of medical treatment.

**10.7** Detailed guidance on compliance with COSHH is given in the Department of Health's (1996) 'Advice on the implementation of the Health & Safety Commission's occupational exposure standards for anaesthetic agents'. Further guidance is given in by the Health & Safety Executive's (1996) 'Anaesthetic agents: controlling exposure under COSHH'.

**10.8** The COSHH regulations set out very specific duties that apply to anaesthetic gases, and employers have a legal obligation to ensure that these duties are discharged. It is therefore the responsibility of the general manager or chief executive to implement the requirements of the COSHH regulations with respect to anaesthetic gases. This subject is covered in Part B.

**10.9** For new installations, an assessment should be made of the transfer and receiving equipment currently in use and intended for use with the new installation. Where the transfer and receiving equipment has been designed to BS 6834:1987, the disposal system design should be to BS 6834:1987. Where the transfer and receiving equipment in use has been designed to BS EN 740:1999, the disposal system should be designed to BS EN 737-2:1998. Where a mixture of equipment is in use, the system should be designed to BS 6834:1987. Where both types of equipment are required to be used on the same disposal system, a restrictor should be provided for the BS EN 740:1999 equipment to restrict the flow to its design flow rate. The system should be installed in all operating departments and other areas, as required, in accordance with the levels of provision given in Table 11.

### Note

BS 6834:1987 covered all aspects of the anaesthetic gas scavenging systems and has now been superseded by BS EN 737-2:1998 and BS EN 737-4:1998. ISO 7396 (in preparation) will replace BS EN 737 Parts 2 and 4.

10.10 A typical system schematic is illustrated in Figure 30 and shows the terminology used.

10.11 The internal components and pipework of AGS disposal systems are in contact with a patient's expired breath. Even though there is considerable dilution by virtue of the receiving system that forms part of the anaesthetic equipment, there is, however, potential for bacteriological

**Figure 30 Schematic diagram of an AGS disposal system**

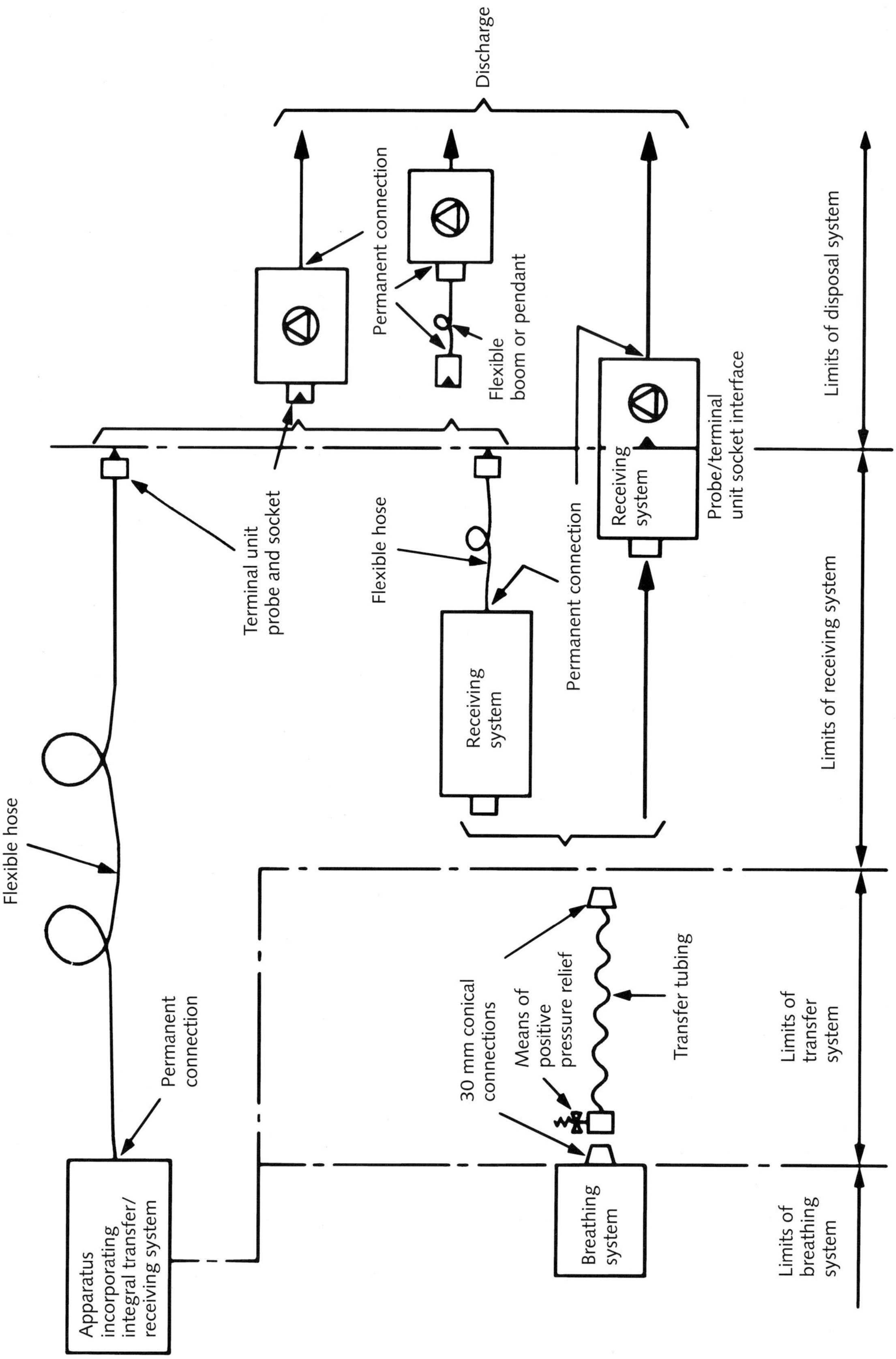

contamination. The materials should be reasonably resistant to corrosion and should withstand cleaning, disinfection or sterilization as appropriate.

10.12 The fixed pipework may be of copper or other suitable material such as PVC. Where copper pipework is installed at the same time as the MGPS, it is desirable to use degreased pipework to the same specification as that used for the MGPS (see Chapter 13) in order to avoid confusion.

10.13 Where PVC pipes larger than 38 mm diameter pass through a fire compartment, they should be protected with metal sleeves extending for 1 m either side of the compartment in accordance with the Building Regulations 2000. The recommendations of Firecode and Health Technical Memorandum 81 should be followed.

## Selecting the number of disposal system pumps

10.14 For operating departments, the number of disposal system pumps should be selected in accordance with the number of air-handling units that are to be installed for each operating suite. For example, if a separate air-handling unit is supplied for each suite, a separate AGS disposal system pump should be installed. Where an air-handling unit supplies two or more operating suites, the AGS disposal system should serve the same number and in this case be a duplex system with automatic changeover. (Where a single pump is provided for an individual operating suite, a spare pump for up to six units should be provided for immediate connection into the system in the event of failure.)

## Flow and diversity

10.15 Although more than one AGS terminal unit may be installed in an operating room or anaesthetic room for convenience, it may be assumed that only one terminal unit in each room will be in use at any given time. The AGS terminal unit in the anaesthetic room and operating room, however, may on rare occasions be in use simultaneously; therefore, the plant is sized for two AGS terminal units for each operating suite.

10.16 The performance criteria for the disposal system are specified in the relevant British, European and International Standards in terms of the extract flows at specified resistance. The disposal system should meet the requirements set out in the table below with the number of terminal units for which it has been designed for use.

| | **Disposal system standard** | | | |
|---|---|---|---|---|
| | **Pressure drop** | | **Flow rate** | |
| | **BS 6834: 1987** | **ISO DIS 7396-2: 2005** | **BS 6834: 1987** | **ISO DIS 7396-2: 2005** |
| Maximum | 1 kPa | 1 kPa | 130 L/min | 80 L/min |
| Minimum | 4 kPa | 2 kPa | 80 L/min | 50 L/min |
| Maximum static pressure | 20 kPa (–ve) | 15 kPa (–ve) | | |

### Notes

Since the preparation of BS 6834:1987, developments in anaesthesiology have resulted in reduced flows being used. Depending on local circumstances, it may be possible to commission systems for different flows in accordance with ISO DIS 7396-2:2005. Details of the test flows should be recorded in the commissioning documentation.

The pump inlet should include a vacuum indicator for commissioning purposes.

## Discharge outlet

10.17 Careful consideration should be given to the siting of the discharge from the disposal system. It should preferably be sited at roof level, well away from ventilation inlets, opening windows and other apertures, to prevent pollution re-entering the building.

## Plant control indication

10.18 There should be indicators to show the following conditions:

a. green "mains on";

b. green "air flow" normal;

c. yellow "duty pump failed" (plant fault);

d. red "system failed" (plant emergency).

10.19 Indicator panels should be installed in operating rooms.

10.20 The "air flow normal" indication should be initiated by either a pressure switch or air flow detection device at the pump.

# 11 Other medical gas pipeline installations

## General

11.1 It is possible to extend medical gas system design concepts to other gases used from cylinders and still maintain the elements of gas specificity that are essential requirements together with all other relevant safety considerations.

## Helium/oxygen mixture

11.2 Helium/oxygen mixture is used by patients with respiratory or airway obstruction and to relieve symptoms and signs associated with respiratory distress. It can be administered by means of face mask and cannula, a demand valve with face mask with cannula attached, a nebuliser, or by a ventilator.

11.3 Its main use will be in Accident & Emergency (A&E), supplied from portable cylinders with integral control valve and regulator, and in critical care areas.

11.4 When provided by means of a pipeline installation, all the elements of a manifold supply system for other medical gases should be installed.

11.5 The manifolds will be designed to operate at low pressure (10 bar), and connection to K-size cylinders will be made by means of a low-pressure flexible assembly to a terminal unit integral with the cylinder regulating valve. The connection to the manifold will be by means of a NIST connector.

11.6 The individual cylinders will include pressure transducers to monitor the pressure upstream of the integral control valve. (Cylinders do not necessarily discharge simultaneously.)

11.7 In the case of helium/oxygen mixture pipelines, the tertiary source of supply is a portable cylinder.

11.8 The manifold should be located close to the facility that it supplies.

| **Compressed gas cylinder manifold systems** | | |
|---|---|---|
| **Primary supply** | **Secondary supply** | **Tertiary supply (third means of supply)** |
| Fully automatic manifold<br>Number of cylinders based on system design | Manual emergency reserve<br>Manifold to come on-line automatically via a non-return valve.<br>Number of cylinders based on ability to provide 4 hours' supply at average use | Locally based cylinders with integral regulator/ flowmeter and terminal unit outlet |

## Oxygen/$CO_2$ mixture

11.9 Oxygen/$CO_2$ mixture has been supplied by pipeline in at least one installation in the UK for anaesthetic purposes in cardiothoracic procedures.

11.10 There has been little interest shown in installing others and, therefore, this medical gas is no longer included within the scope of this Health Technical Memorandum.

## Carbon dioxide

11.11 Carbon dioxide is now not generally used as a respiratory stimulant post-operatively. Pipelines have not been installed in the UK for respiratory applications. Its main use today is for insufflation during surgery, and to date there have been some installations in the UK.

11.12 When pipeline systems are installed for such purposes, the general requirements for other medical gas pipelines should be followed. The terminal unit should comprise a NIST connector with integral check valve contained in the surgeon's pendant. The level of provision of AVSUs should be provided as for other medical gas pipelines.

11.13 A semi-automatic manifold will normally be satisfactory and it should be installed "locally".

A 2 x 4 VF-size manifold will provide adequate capacity. The safety valve discharge should be taken outside the department. The warning and alarm system indicator will normally be installed in the operating room control panel.

## Nitric oxide

11.14 Treatment using nitric oxide is subject to specific Ph. Eur. requirements. Distribution of the gas by pipeline systems is not considered appropriate.

# 12 Warning and alarm systems

## General

12.1 The provision of a warning and alarm system is essential to monitor the safe and efficient operation of an MGPS. There are three reasons for this monitoring:

a. to indicate normal function of the pipeline system by means of visual indicators;

b. to warn by visual and audible indication that routine replacement of cylinders or other engineering action is required;

c. to inform the user by visual and audible emergency alarms that abnormal conditions have occurred which may require urgent action by the user. This alarm condition will require a rapid response by the various departmental staff.

12.2 To date, practice has been to have a "dedicated" medical gas warning and alarm system and this approach will remain in many situations. With the development of computer-based integrated patient/management systems, nurse call and other alarm systems, however, there is considerable scope for including medical gas system information including text action prompts etc. Additionally, building management IT-based systems will play an increasing role in the operation and management of an MGPS.

## Dedicated systems

12.3 The requirements of "dedicated" warning and alarm systems are covered in paragraphs 12.3 to 12.62 and a schematic diagram of a typical system layout is shown in Figures 31 and 32. Warning and alarm systems are required for all medical gas and vacuum systems. A simplified system is required for surgical air systems and for the AGSS, with the warning/indication panel located in the operating room.

12.4 Warning and alarm systems comprise pressure sensors, a central system providing information on all monitored functions, with repeater panels located where information is required to ensure the necessary action is taken. Area alarms should be provided to give warning to users downstream of the designated departmental AVSU (see Chapter 3).

12.5 Pressure sensors should be connected to the pipeline by means of minimum leak devices.

12.6 All MGPS warning and alarm indicating panels should comply with the requirements of this Health Technical Memorandum, including all operating room panels.

## Panel location

### Central indicator panel

12.7 Warning and alarm conditions for all medical gas supply systems should be displayed on a central panel located in a position where there is continuous 24-hour occupation, such as the telephone switchboard room or the porter's lodge.

### Repeater indicator panel location

12.8 Repeater panels should be provided in other locations to display all or some of the information on the central alarm so that appropriate action can be taken to ensure the continuing operation of the system. Some warning system information may be appropriate for display in specific departments, for example cylinder manifold status information in a porters' room, and oxygen concentration in the pharmacy department when a PSA plant supplies the hospital pipeline installation.

### Area warning and alarm panel location

12.9 Local panels to display "high" and "low" gas pressure should be installed in the locations given in Chapter 3. The sensors for these panels should be located downstream of the designated AVSUs, normally the departmental AVSUs. It should not be possible to isolate the sensor with a separate shut-off valve and they should be connected to the pipeline by means of a minimum leak device.

Figure 31 Typical warning and alarm system layout (reproduced by kind permission of Shire Controls)

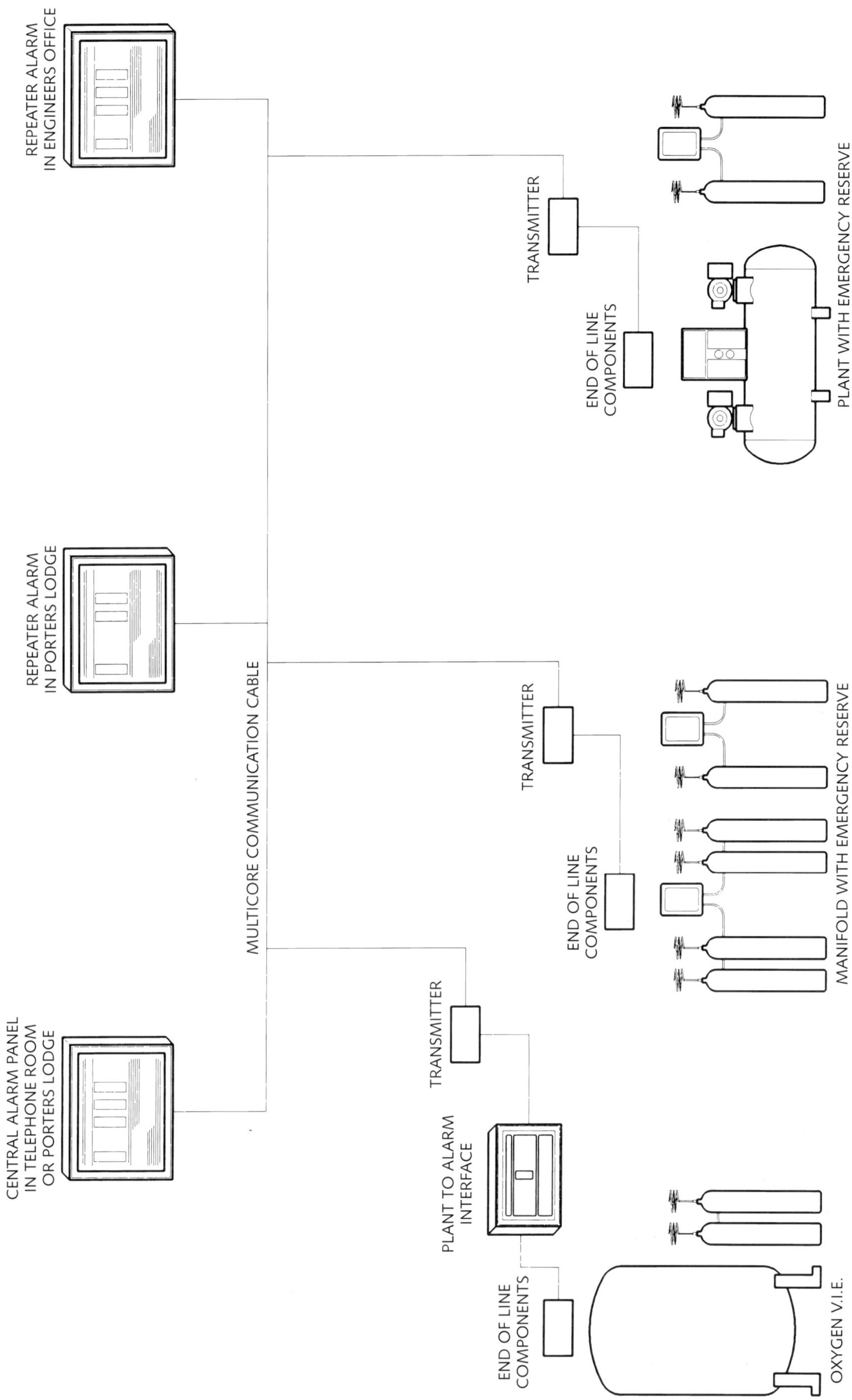

**Figure 32 Typical area alarm panel (reproduced by kind permission of Shire Controls)**

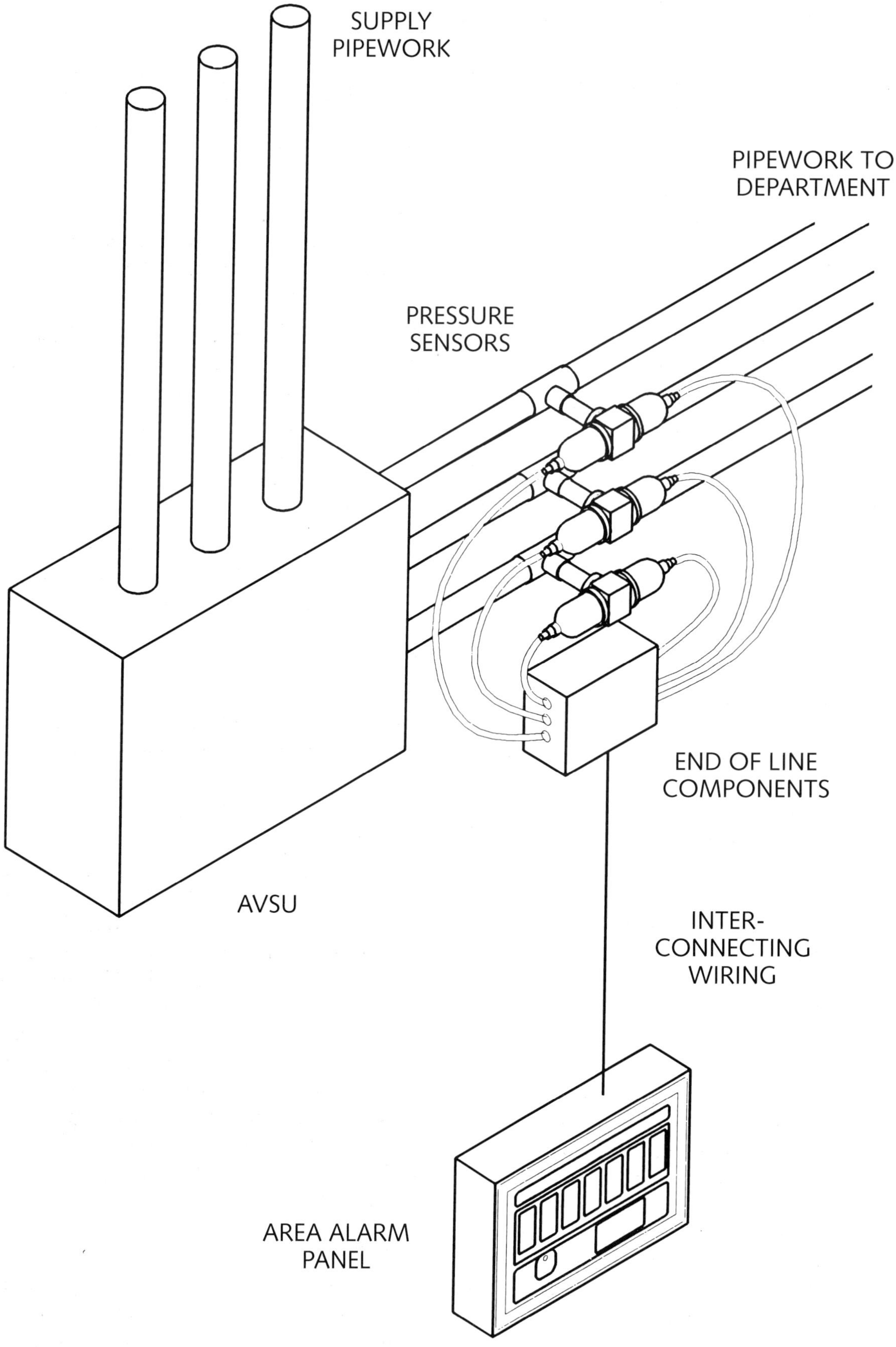

## System components

12.10 Warning and alarm systems include the following functional elements:

a. interfaces/transmitters that convert the signal from the plant or manifold volt-free alarm contacts into a form which can be transmitted via multiplexed cable (for example using pulse-width modulation). The transmitter may be a separate unit or may be incorporated:

(i) in plant or into a manifold control panel;

(ii) into an indicator panel.

Cases (i) and (ii) should include line-fault monitoring devices;

b. indicator panels which display the transmitted signals;

c. interconnecting multiplex wiring which connects all interfaces/transmitters to all indicator panels.

## System layout

### Central system

12.11 A typical system layout is shown in Figure 31, which shows initiating devices at remote locations such as the VIE compound, medical air and vacuum plantrooms, nitrous oxide manifold room and emergency/reserve manifold rooms. The transmitters are normally located close to the initiating devices. Indicator panels are typically located at the telephone exchange, the porter's room and the engineer's office to provide information requiring action by engineering and other support staff.

### Area warning and alarm systems

12.12 A typical layout of an area warning and alarm system is shown in Figure 32. For each gas service there should be local pressure switches for low pressure; high pressure switches are also required when oxygen, nitrous oxide and medical air are installed together. These conditions should be indicated on a locally-mounted indicator panel, with facility to provide a common alarm condition for connection to other alarm panels. Area panels carry no indication of the warnings for cylinder replacement and plant functions that are given on central indicator panels.

## General requirements

### Labelling

12.13 All visual signal panels should be permanently labelled according to their function, including clear identification of the areas, rooms or departments served.

### Visual signals

12.14 Flashing visual signals should have alternate "on" and "off" periods, each of equal duration between 0.25 and 0.50 seconds.

12.15 There should be two separately energised light sources for each signal, arranged so that the failure of one source does not affect the other.

12.16 The light sources should have a design life of at least five years of continuous operation.

### Audible signals

12.17 All audible signal tones should be modulated equally at a rate of 4 Hz ±10% between two tones of 440 Hz ±10% and 880 Hz ±10%.

### Automatic resetting

12.18 When a warning or alarm signal occurs and the system condition subsequently reverts to normal, the corresponding visual and audible signals should automatically reset to normal.

### Temporary muting

12.19 Means must be provided on each panel for the user to mute the audible signal. The signal must re-sound after a nominal 15-minute period if the fault condition still exists. The process of muting and reinstatement of the signal should be repeated until the fault condition has been rectified. Operation of the mute on the central panel should be accompanied by change from flashing to steady illumination of the corresponding visual indicator on the central and any repeater panels. Operation of the mute on area alarm or repeater panels should not be accompanied by a change from flashing to steady illumination.

### Continuous muting

12.20 An internally-mounted switch should be provided to allow continuous muting during periods of maintenance. When the system condition returns to normal, the continuous muting should automatically reset to normal operation. When the continuous muting is in operation on any alarm condition, it should not prevent the operation of the audible signal on other alarm conditions when a fault condition arises.

### Electrical wiring

12.21 All electrical wiring should be in accordance with IEE regulations.

### System integrity

12.22 If extra low voltage (ELV), maximum 50 V, is superimposed on the signal or communication circuit (for example by cross-connection), the system design should ensure that any damage to the system is limited to replaceable panel components and that such damage is indicated as a system fault.

12.23 The performance of the system should not be compromised by the use of multi-core cabling that carries ELV and communication signals in adjacent cores.

12.24 The system should be designed to reject spurious radio frequency (RF) or mains noise typically arising in hospitals, examples being diathermy equipment and current spikes caused by plant start-up etc.

### Relay conditions

12.25 If relays are used to transmit alarm signals, the relays should be energised in their normal closed condition.

### Mains power supply

12.26 The mains electricity supply should be derived from the essential power supply (that is, must be on the emergency system).

### Safety extra low voltage/functional extra low voltage power supply

12.27 The panel power may be designed either as a safety extra low voltage (SELV) system or as a functional extra low voltage (FELV) system, as defined in Part 4 of the IEE Wiring Regulations.

12.28 The ELV power supply may be housed either in the alarm panels or in a separate metal enclosure.

12.29 The power supply should be rated for the full load of the panel, with visual and auditory signals on all normal and alarm conditions.

### Test facility

12.30 Each panel should be provided with a means to test all visual and audible signals on that panel. The power supply should be capable of sustaining all indicators and audible signals.

## Warning and alarm system faults

### General

12.31 A flashing red visual indicator and an audible signal should operate on all panels when any of the following conditions occur:

a. line fault from the initiating device;

b. communication fault or other wiring fault;

c. mains power failure.

### Line fault

12.32 The system should monitor the integrity of the lines between the initiating devices and the panel or transmitter units. The "alarm system fault" condition should be indicated on loss of integrity, for example open or short circuits, together with the visual alarm indicator(s) associated with the faulty wiring.

### Communication/wiring fault

12.33 The system should indicate an alarm system fault in the event of loss of data transmission between panels and transmitters.

### Mains power failure

12.34 Failure of mains power should be shown by a flashing red indicator and an audible signal, which should be powered from an internal battery. The audible signal may be muted and not automatically reinstate as required under normal power supply (see paragraph 12.19), but the visual indicator should continue to flash until either the fault has been rectified or the battery has discharged.

### Stand-by battery

12.35 A battery should be provided with sufficient capacity to power the visual and audible "alarm system fault" signal for a minimum period of four hours. The battery should be sealed and exchangeable, and should automatically recharge within 72 hours.

#### Legend

12.36 The legend on this indicator should be "alarm system fault".

## Indicator panel requirements for all systems

#### Indicators

12.37 Panels should be provided with all indicators for the gas services in local use.

12.38 The visual indicators should be arranged vertically in priority order, with the normal indicators at the top. The sequence of gas services should be, from left to right:

a. medical oxygen (cryogenic and cylinders/ pressure swing adsorber (PSA) systems);

b. nitrous oxide;

c. nitrous oxide/oxygen mixture;

d. medical air 400 kPa (compressor plant, cylinders and synthetic air);

e. surgical air 700 kPa;

f. medical vacuum (pumps);

g. helium/oxygen mixture.

12.39 In addition to the gas service signal indicators, each panel must include:

a. a green "power on" indicator without an audible signal;

b. a red "alarm system fault" indicator with an audible signal.

#### Labelling

12.40 Panels should be labelled as follows:

a. medical gas alarm;

b. with the identification of the medical gas services indicated, and the areas and departments served.

#### Construction

12.41 The fascia panel should be removable to allow access to the rear of the fascia or to the panel for maintenance purposes.

12.42 Access to the interior of the panel should be tamper-proof.

12.43 It should be possible to replace the source of illumination without removing the legend.

12.44 Panels should have electrical sections with protection at least equal to BS EN 60529:1992.

12.45 Panels and their housings should be of adequate strength for their purposes and be manufactured from corrosion-resistant materials.

12.46 If gas services are brought into the panel, they should be housed in separate, enclosed compartments, which are vented to the outside.

12.47 There should be gas-tight seals where electrical services pass through any gas compartment.

#### Remote audible sounder

12.48 All panels should have provision for connection to a remote audible sounder.

## Central indicator panel requirements

#### Displays

12.49 The central panel should display all signals for all MGPS which are generated by the warning and alarm system, as described in paragraphs 12.50–12.53.

*Normal*

12.50 The normal condition for all piped MGPS should be displayed as a steady green visual signal. The "normal" indicator should extinguish in warning and alarm conditions.

*Warnings*

12.51 Warning conditions appropriate to each MGPS should be displayed as a flashing yellow visual signal that may be accompanied by a mutable audible signal (see Table 24).

*Emergency alarms*

12.52 Emergency alarms are generated by loss of pipeline pressure or vacuum and are indicated by flashing red visual signals accompanied by mutable audible signals.

*Alarm system fault*

12.53 The "alarm system fault" condition should be displayed as a flashing red visual signal accompanied by a mutable audible signal.

#### Mute functions

12.54 The temporary mute should cancel the audible signal for about 15 minutes and change the visual indicators from flashing to continuous on all central and repeater panels.

12.55 Operation of the continuous mute should inhibit the 15-minute reinstatement of the audible alarm.

12.56 Operation of the mute should not inhibit the visual or audible indication of any subsequent alarm conditions.

### Panel legend and display

12.57 Panel legend and display should be as shown in Table 24.

## Repeater indicator panel requirements

### Displays

12.58 The repeater indicator panel should always display "normal", "emergency alarm" and "alarm system fault" conditions as given above. The repeater panel should display some or all of the warning conditions that are displayed on the central indicator panel. The extent of the display of warnings should be varied to suit local clinical requirements.

### Mute functions

12.59 The temporary mute should cancel the audible signal for about 15 minutes whilst the visual indicator continues to flash. Operation of the temporary mute (on the central panel) should change the visual indicator to continuous illumination on the central and any repeater panels.

12.60 Operation of the continuous mute must inhibit the 15-minute reinstatement of the audible alarm.

12.61 Operation of the mute should not inhibit the visual or audible indication of any subsequent alarm conditions.

### Panel legend and display

12.62 The panel legend and display should be as shown in Table 24.

## Area warning and alarm panel

### Panel displays and legend

12.63 Area panels should display the conditions listed in Table 25.

### Mute functions

12.64 The temporary mute should cancel the audible signal for about 15 minutes whilst the visual indicator continues to flash.

12.65 Operation of the mute should not inhibit the visual or audible indication of any subsequent alarm conditions.

## Integrated systems

12.66 The introduction of computer-based systems for a range of functions such as patient information, nurse call and other alarm conditions provides an opportunity to further include certain provisions of medical gas pipeline warning and alarm conditions. This concept is totally new and, at this stage, the applications have not been thoroughly evaluated or analysed. One of the advantages of the concept is that text prompts can be displayed on the computer display when changes in the status of the pipeline occur, and these prompts can advise staff of the need to take specific action.

12.67 The advantage of a computer-based system is that the advice given in the text message can be varied to take account of specific circumstances, changes in operating procedures and functional changes within individual departments. Such systems are likely to be of most use in in-patient ward accommodation; it may not be appropriate for central warning and alarm conditions or in individual operating rooms and other accommodation in which anaesthetic procedures are taking place.

12.68 It will be necessary to change the perception of users in that with this approach the "normal" conditions of the pipeline systems that are continuously displayed on alarm indicator panels will not exist – audible emissions and displayed messages generated by the computer-based system will be in response to changes from the "normal" situation. To ensure the long-term viability of the system, any supplier or installer of such a system must supply sufficient information about the system to allow modification, expansion or replacement of sections of the system by a third party. This must include source code for any software, passwords and details of any other security device, and details of any communication protocols. This information must be handed to the end-user before the system is accepted by the end-user.

### Note

No further information can be given at this stage until further development and consultation takes place.

## Table 24 Signals and displays for central alarm panels and repeater panels

| Supply system[1] | Alarm conditions | Legend | Colour | Audible system | Location[2] |
|---|---|---|---|---|---|
| Automatic manifolds | 1. Duty bank empty: stand-by bank running | Change cylinders | Yellow | Yes | A B |
| | 2. Stand-by bank below 10% capacity | Change cylinders immediately | Yellow | Yes | A B |
| Compressed cylinders on automatic manifold serving a single vessel cryogenic oxygen system or liquid cylinder installation | Changeover of manifold. Pressure in each bank is not monitored. | Reserve low | Yellow | Yes | A B |
| Compressed cylinders on reserve manifold serving a compressor plant | Changeover of manifold. Pressure in each bank is not monitored. | Reserve low | Yellow | Yes | A B |
| Compressed cylinders on reserve manifold serving an automatic manifold | Reserve pressure below 68 bar (<14 bar for $N_2O$) | Reserve low | Yellow | Optional | A B |
| Medical air compressor and surgical air compressor | 1. Plant fault | Plant fault | Yellow | Yes | A B |
| | 2. Plant emergency | Plant emergency | Yellow | Yes | A B |
| Medical vacuum plant | 1. Plant fault | Plant fault | Yellow | Yes | A B |
| | 2. Plant emergency | Plant emergency | Yellow | Yes | A B |
| Oxygen concentrator | 1. Plant fault | Plant fault | Yellow | Yes | A B |
| | 2. Plant emergency | Plant emergency | Yellow | Yes | A B |
| Pressure fault (pipeline) high or low and oxygen concentration fault for PSA plant | For each gas service to indicate that the pressure in the distribution system has risen/ fallen from the "normal" working pressure given in Chapter 4 and, for PSA plant, that $O_2$ concentration <94% | Pressure fault | Red | Yes | A B |
| Vacuum pressure (pipeline) | To indicate that vacuum in the pipeline distribution system has risen above the normal working pressure given in Chapter 4 | Pressure fault | Red | Yes | A B |

**Notes:**

1. For liquid supply systems, see Chapter 6.
2. A = Central indicator panel – telephone room and/or porters' room, ie 24-hour occupancy.
   B = Facilities management office reception.

## Table 25 Area panel legend and display

| Alarm function | Legend | Colour | Auditory signal |
|---|---|---|---|
| For oxygen, nitrous oxide and medical air[1] to indicate that the pressure in the pipeline serving the department has risen above the normal value given in Chapter 4 | High pressure | Red | Yes |
| For each gas service to indicate that the pressure in the pipeline serving the department has fallen below the normal value given in Chapter 4 | Low pressure | Red | Yes |
| For vacuum to indicate that the pressure in the pipeline serving the department has risen above the normal value given in Chapter 4 | Vacuum fault | Red | Yes |

**Notes:**

1. A high pressure alarm is only required when oxygen, nitrous oxide and medical air are installed together.

For location of area panels, see Table 11.

# 13 Pipeline installation

13.1 Generally, MGPS should be kept away from areas where they may be subject to any of the following:

a. mechanical damage;

b. chemical damage;

c. excessive heat;

d. splashing, dripping or permanent contact with oil, grease or bituminous compounds, electrical sparks etc.

13.2 Service ducts or voids containing pipelines that include valves etc should have adequate ventilation to prevent gas build-up in the event of any leakage. Elsewhere, where pipelines are brazed throughout their entire length (and where they will have been subjected to a pressure test), no ventilation is required.

13.3 Exposed pipelines should not be installed in lift shafts, kitchens, laundries, boilerhouses, generator rooms, incinerator rooms, storage rooms designed to house combustible materials, or in any other fire-risk areas. Where pipelines in hazardous areas are unavoidable, they should be enclosed in non-combustible, non-corrosive materials that have no electrolytic reaction with copper in order to prevent the possibility of the liberation of gases into the room in the event of pipeline failure. Medical gas pipelines should be routed away from natural gas pipelines where there is a potential for a flammable gas mixture to accumulate in the case of a leak.

13.4 Where pipelines are run in enclosed ducts with other services such as steam mains and water supply systems, they should be inspected regularly as corrosion can occur as a result of chloride deposits following leakage. They should not be run in enclosed ducts with other services where they cannot be inspected.

13.5 External pipe runs should be avoided when possible. Where external runs, however, are necessary, they should be protected as follows:

a. **on external vertical surfaces up to the maximum height of exposure to possible damage (for example vehicular movement):** by means of galvanised, profile-section steel of sufficient thickness to afford adequate protection. The protection should cover the entire space taken up by the pipeline(s), but stand off the surface such that the pipes can be inspected visually (The armour should be readily detachable to permit more detailed inspection.);

b. **when crossing horizontal surfaces, roofs etc:** similar protection to (a) above should be provided to withstand "stepping" damage using profiled section, as above.

13.6 Pipework should be protected from lightning strikes by ensuring that they are run within a 60-degree cone beneath the lightning conductor, for example when run along parapet walls, or when penetrating parapet walls. When run across roof surfaces, a copper lightning conductor should be run on the top surface of the pipework cover providing physical protection, and should be bonded to it.

13.7 Internal pipelines should be suitably protected where there is a possibility of physical damage, for example from the passage of trolleys, tugs etc.

13.8 Wherever practicable, a clearance of at least 25 mm should be maintained between each service and 150 mm should be the separation distance between the medical gas pipeline and heating pipes, hot water service and steam pipelines. Where pipelines cross over other services and a clearance of 25 mm cannot be maintained, they should be electrically bonded and wrap-insulated, in accordance with IEE regulations. They should be bonded to main earth at building entry and exit. Care is required when selecting pipeline routes to prevent the pipes coming into contact with electric cables and wiring, and to minimise the risk of electric shock in the event of a fault on adjacent cables (see Chapter 2).

13.9 Underground pipelines should be run in properly drained ducts not less than 450 mm x 450 mm which have removable covers. Where it is not possible to provide removable covers, two pipes should be run in separate trenches with valves provided in a convenient location at either end. The valves should comprise LVAs with NIST connectors for the purposes of pressure and other tests. The separation distance between the two trenches should be not less than 2 m (see Figure 33). The two pipes should each be sized for the design flow. One or more different gas pipelines can be run in each trench. The route of the pipeline should be identified on the surface and should be clearly shown on site layout drawings. The possibility of installing a "ring-main" (see Figure 34) or double-end supply should also be considered for both air and oxygen within the curtilage of the building.

13.10 Pipelines concealed within walls should have their route clearly shown on "as-fitted" drawings. Pipelines should not be encapsulated in floors, and any joints should be kept to the minimum practicable. Pipelines in stud or plasterboard walls or partitions are acceptable, but the pipeline should be protected from corrosion. If the enclosure of pipelines within plaster wall finishes is unavoidable, they should be wrapped in protective grease-free tape.

13.11 Pipelines need further protection in certain circumstances as follows:

a. where pipes pass through walls, partitions or floors, they should be provided with sleeves of copper pipe (with fire stopping) and, where exposed to general view, be provided with appropriate wall or ceiling plates;

## Figure 34 Typical ring-main arrangement

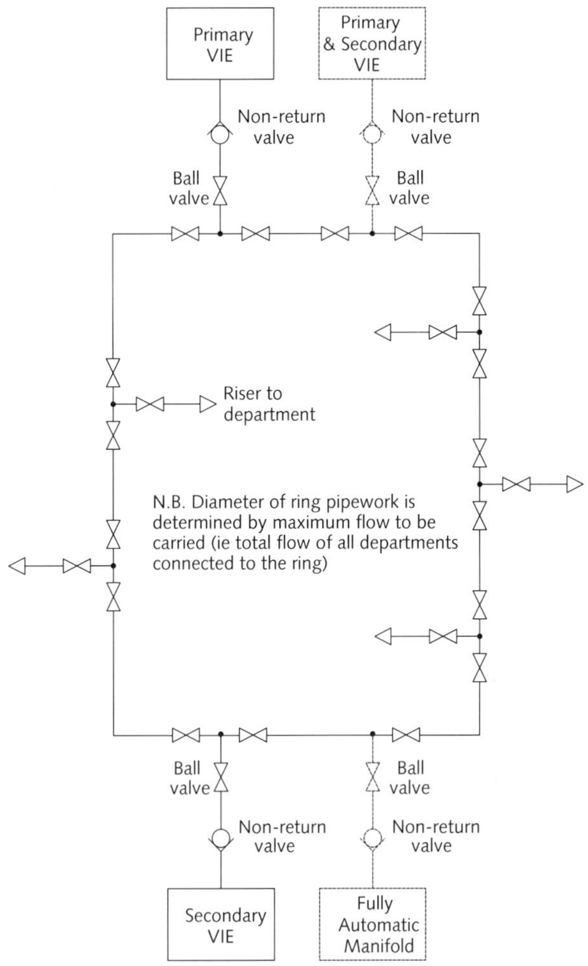


This diagram shows a single feed from each source to the ring main. However, attention is drawn to paragraph 13.9, which details the provision of double feeds to reduce the risk of supply failure arising from mechanical damage to vulnerable pipework.
Note: NIST connectors should be fitted either side of each valve. Where three valves are installed close together, only one needs to be installed between the set of three.

## Figure 33 Typical twin pipeline supply arrangement

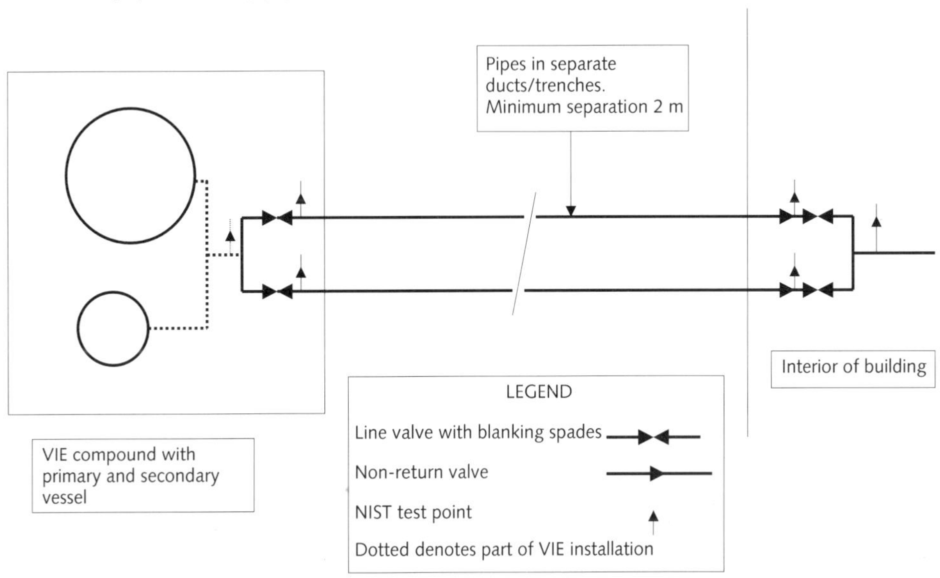

b. in radiodiagnostic procedure rooms etc, radio frequency (RF) screening wave guides may be required (the advice of the equipment manufacturer should be sought);

c. corrosion of pipes can occur where they are in contact with timber that has been treated with fire-resistant or flame-retardant compounds, for example some timber used for roof trusses and floor joists.

13.12 This contact should be avoided by the use of impermeable non-metallic materials in the area where contact may occur. PVC spacers or adhesive PVC tape may be used for this purpose. If spacers are used they should not be liable to drop out due to shrinkage or subsequent movement of the pipe or timber.

13.13 Such precautions are not required where untreated timber is used or where the treated timber is effectively sealed with paint or varnish before the pipes are fixed to it.

## Pipeline materials

### Quality

13.14 The manufacturer should comply with BS EN ISO 9001:2000 for pipes and for all materials including fittings, terminal units etc. A complete specification is given in Model Engineering Specification C11 – 'Medical gases'.

13.15 Where materials are obtained from suppliers from other countries, the suppliers should be registered in accordance with BS EN ISO 9001:2000.

### Pipes

13.16 Material for pipes should be manufactured from phosphorus deoxidised, non-arsenical copper to BS EN 1412:1996 grade CW024A (Cu-DHP) in metric outside diameters and to:

- BS EN 13348:2001 – R250 (half hard) for sizes up to 54 mm; or
- BS EN 13348:2001 – R220 (annealed) for larger sizes.

Stainless steel is a suitable material for medical gas pipeline installations, but is not currently used by installers, and Standards have yet to be established.

### Pipe jointing fittings

13.17 In addition to the above, pipe jointing fittings should be end-feed capillary fittings to BS EN 1254-1:1998.

> **Note**
>
> For straight couplings, expanded joints may be used.

### Other fittings

13.18 Other fittings for connection to copper pipes (for example valve and control panel fittings) may be of copper, brass, gun-metal, bronze or stainless steel.

## Cleaning

### Pipes

13.19 All pipes must be cleaned and degreased for oxygen service and be free of particulate matter and toxic residues in accordance with BS EN 13348:2001. They must be individually capped at both ends and delivered to site identified as medical gas pipes.

### Pipe jointing fittings

13.20 All pipe jointing fittings and sub-assemblies of fittings for connection to pipes must be cleaned and degreased for oxygen service and be free of particulate matter and toxic residues. They must be individually sealed in bags or boxes and delivered to site identified as medical gas fittings.

13.21 Although it is not essential to degrease vacuum installations, these are frequently installed by the contractor simultaneously with the medical gas pipelines. Degreased pipe and fittings should therefore be used for the vacuum installations to avoid confusion. PVC pipework may also be used for vacuum and AGSS but is unlikely to be of benefit other than for exhaust discharges.

> **Note**
>
> Pipes should only be cut with wheel pipe cutters, not hacksaws, to prevent the ingress of copper particles.

## Pipeline jointing

### General

13.22 Except for mechanical joints, only copper-to-copper joints will be permitted on site, made with brazing filler rods that can be used without flux.

## Note

Brazing is performed at a higher temperature than in the case of silver soldering with capillary fittings; the exterior of the pipe will therefore have considerably darker oxide deposits.

13.23 Copper joints to brass or gun-metal fittings will require the use of flux, with subsequent cleaning to remove the flux residues and oxide deposits.

13.24 Heating of the joint for brazing should be carried out with oxygen/acetylene or acetylene, liquid petroleum gas/oxygen torches. Additional heating may be required for some fittings, for example, by means of a second torch.

13.25 The techniques recommended cover all copper-to-copper joints and all copper-to-brass/gun-metal/bronze joints in an MGPS, and are explained in more detail below.

13.26 The brazing technique should be used on all medical gas pipeline services.

### Pipe preparation

13.27 Pipe ends should be cut square with the pipe axis, using sharp wheel cutters whenever possible, and be cleaned to get rid of any cuttings or burrs. Expanded joints should be made using the appropriate tools and dies. Only where the cut pipe has either deformation or a burr which significantly restricts the flow of gas will de-burring be necessary. Only oil- and grease-free tools and dies should be used.

13.28 When brazing copper-to-copper joints:

a. the brazed joints should be made using a silver-copper-phosphorus brazing alloy CP104 to BS EN 1044:1999. No flux should be used;

b. ensure adequate protection of adjacent pipe runs and other services.

## Note

Brazing copper to brass/gun-metal/bronze is not performed on site. Manufacturers use copper-silver-zinc brazing alloy AG203 to BS EN 1044:1999 with an appropriate flux. The flux residues created by the process are chemically removed and, if necessary, the complete assembly is cleaned and degreased for oxygen service. Where brass/gun-metal/bronze fittings are required to be installed they should be supplied complete with copper "tails" of adequate length to ensure that the brazing process does not damage the components.

### Use of $N_2$ internal inert gas shield

13.29 Brazing should be carried out using oxygen-free nitrogen as an internal inert gas shield to prevent the formation of oxides on the inside of the pipes and fittings. This method leaves a bright, clean bore. Some slight burnishing may occasionally be observed on sectioned joints. Purging is still required to remove the internal shield gas and the other particulate matter not associated with the brazing operation.

13.30 Oxygen-free nitrogen should be supplied to the inside of the pre-assembled, unbrazed pipework through a pressure regulator and flow controller or flow-regulating device.

#### *Application*

13.31 Oxygen-free nitrogen as an internal inert gas shield should be used for all positive pressure gases and for vacuum pipelines – up to and including 22 mm – that are run in medical gas supply units and to individual terminal-unit drops. Once the route of the vacuum service has been clearly established (that is, above the ceiling level), nitrogen purging may be discontinued. Nitrogen purging is not required for AGS disposal systems.

## Note

During the first-fix stage of pipeline installation, particularly when installing in confined locations such as medical supply units or running pipework within partitions etc to individual terminal unit drops, it is possible to inadvertently crossover a pipeline. This is usually discovered at an early stage and, so that the pipe section can be re-assigned and the fault can be corrected, it is essential to use the shield gas to maintain the cleanliness of the internal bore.

13.32 By agreement between the health facility management and the pipeline contractor, the use of a purge gas may be waived on joints such as break-ins to old pipeline systems, where pipe joints will not have been made in accordance with this technique.

13.33 It is recommended that the pipeline to be brazed should first be flushed to remove the air. This may be followed during the brazing operating by a continuous or intermittent flow as necessary to prevent the ingress of air. Pipe ends may be capped if desired to direct the flow of nitrogen into sections of the pipe or pipes to be brazed. Particular attention should be given to the gas shielding of T-joint fittings. Care should also be taken to ensure that other pipelines in close proximity to the one being brazed do not oxidise due to heat transfer. It is essential that there is a leak-free connection between the pipework to be brazed and the nitrogen supply.

*Safety*

13.34 If working for prolonged periods in very confined spaces, precautions must be taken to avoid excessive build-up of nitrogen by ventilating the space or by piping the shield gas safely out of the space. The oxygen content of the ambient air should be monitored when brazing in a confined space.

*Control of cylinders*

13.35 The contractor and the site engineer must keep a record of nitrogen cylinders held on a site. Nitrogen cylinders should be accounted for and removed from the site at the end of the contract, and must not become mixed up with medical gas cylinders.

## Inspection of joints

13.36 Inspection of joints should be carried as a "rolling" procedure on a monthly basis as work progresses for each team performing the installation in accordance with the following procedure:

a. the site engineer should identify a number of fittings to be cut out for examination in order to establish the quality of the finished joint. The exact number to be cut out will vary with the size of the installation: as a guide, a ratio of one fitting per 200 should be cut out; a minimum of ten for all systems should be cut out for examination (it is preferable to perform these checks before pressure-testing sections of pipeline);

b. the fittings cut out should be cut open (quartered longitudinally) and examined. If unacceptable joints are found, adjacent fittings should be cut out until the extent of any faulty workmanship has been established. This may require extensive removal of sections of the installation.

### Internal cleanliness

13.37 The tube and fitting should be internally clean and free from oxides and particulate matter. Some heat burnishing may be apparent and is acceptable.

### Penetration

13.38 Penetration of brazing alloy:

a. due to tolerances of the capillary space on these pipes and fittings, full penetration of the brazing alloy may not occur and is not necessary;

b. the minimum penetration at any point on the joint must be three times the wall thickness of the tube or 3 mm, whichever is greater;

c. the pipe should be fully inserted up to the shoulder of the fitting.

**Note**

These tests can be carried out on a sectional basis.

## Jointing methods (mechanical)

13.39 In addition to mechanical connections to plant and valve assemblies, mechanical connections can be used for connecting pre-piped bed-head trunking/wall units to the pipeline distribution system. They may also be used in situations where brazing may present a fire risk and in other situations when patients cannot be transferred to alternative accommodation (in which case they should be of the permanently swaged type). Mechanical connections should have comparable structural integrity to brazed fittings in normal operation and in the event of fire. Any lubricant required for swaging should be oxygen-compatible. The fittings should not contain elastomeric materials.

13.40 Mechanical joints can also be used in an emergency. Fittings that are non-permanent should be number-tagged and marked on the record drawings.

**Note**

Open ends of the remaining pipework should be capped off. The installation should be made good as soon as possible in accordance with paragraphs 13.29–13.35.

13.41 PTFE tape is not an acceptable sealing material on oxygen systems or elsewhere downstream of final filters on supply plants.

## Note

PTFE tape, if applied, can enter the gas system and fragments can block terminal units and present a fire hazard with high-pressure oxygen. Also, when applied by hand, traces of oil and grease can contaminate the inside of the pipeline.

13.42 Liquid or gel-sealing media should be used only if they have been tested and proven safe when subjected to the tests specified in BS EN ISO 15001:2004.

## Capping

13.43 Sections of pipeline should be capped as soon as they are completed so as to prevent the ingress of debris.

## Pipe supports

13.44 The pipeline should be adequately supported at sufficient intervals in accordance with Table 26 to prevent sagging or distortion. Supports for surface-mounted pipework should provide clearance to permit painting of the surface. Where it is essential for pipes to cross electric cables or conduit, they should be supported at intervals on either side of the crossing to prevent them from touching the cables or conduit. Supports should be of suitable material or suitably treated to minimise corrosion and prevent electrolytic reaction between pipes and supports.

### Table 26 Intervals between copper pipe supports (horizontal and vertical)

| Outside diameter (mm) | Maximum interval between supports (m) |
|---|---|
| Up to 15 | 1.5 |
| 22–28 | 2.0 |
| 35–54 | 2.5 |
| >54 | 3.0 |

**Note:** Consideration should be given to additional supports near LVAs, elbows etc where the potential effects of inadvertently applied torque can result in severe pipeline distortion or fracture.

13.45 Pipelines need not be laid with falls. In the case of vacuum, the sub-atmospheric pressure will result in the evaporation of any moisture entering the system.

13.46 The connection of individual, or a number of, vacuum terminal units into branches should be taken into the top of the pipeline to avoid flooding other vertical pipe drops, should liquid carry-over occur. Each vacuum main riser should be provided with a double (15 mm) valve arrangement at the base, with an intervening full bore pipe section, preferably transparent, to permit drainage when the system is under vacuum; one of the valves should be lockable in the closed position. Within trunking systems and medical supply units etc, vacuum pipes should connect into the underside of terminal units.

## Identification of pipelines

13.47 Pipelines should be identified in accordance with BS 1710:1984, and colour banding for the pipelines should be used. Colour band identification (see Figure 35) should be applied near to valves, junctions, walls etc. A label applied every 3 m and bearing 6 mm size letters should identify each gas. Self-adhesive plastic labels of

### Figure 35 Pipeline identification colours

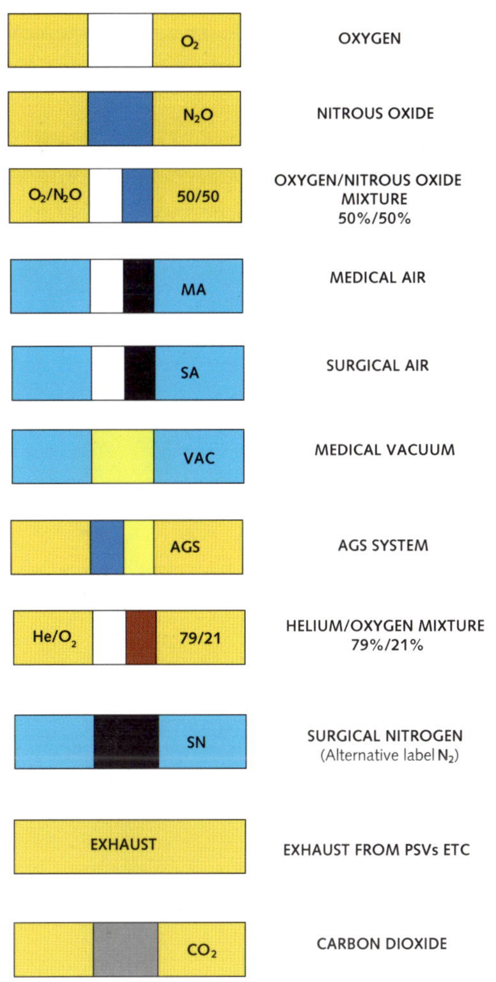

approved manufacture may be used for this purpose. A band 150 mm wide is usually adequate. All colour-coded tapes applied by the pipe manufacturers should be removed before the systems are identified, in accordance with this paragraph.

13.48 Care should be taken to maintain pipeline identification when periodical re-painting is undertaken. The direction of flow should be indicated.

## Pipeline components

13.49 Pipeline components, which may be attached to an MGPS, include various types of terminal unit, AVSUs and other components such as emergency inlet ports and pressure control equipment.

## Medical supply units

13.50 These should comply with BS EN ISO 11197: 2004. The construction should provide segregation of FELV electrical services by means of partitions or flexible conduit as appropriate. Access to "live" components should be via panels that are removable by means of tools only. Multi-purpose medical supply units should be constructed in such a manner to ensure that flexible hoses are not subject to excessive "kinking" or "twisting". The flexible hose materials should be free from volatile or organic compounds and be tested prior to installation. Rigid units should be piped in copper.

13.51 When these fittings include flexible connecting assemblies for the gas supply, the method of attachment to rigid pipework or terminal units should be by means of the appropriate NIST connector in accordance with BS EN 737-1:1998.

13.52 The fittings should be provided with adequate venting to allow escape of gas in the event of rupture of one or all of the medical gas services.

13.53 The recommended height for rigid pendants is 2000 mm above FFL. The maximum height for pendants capable of vertical movement should be 2000 mm above FFL in the fully retracted position.

13.54 The use of medical air for pneumatically actuated pendants is covered in Chapter 3.

### Note

In cases where medical or surgical air terminal units are not required to be included in these pendants, an AVSU will still be required "locally" for emergency isolation and servicing of the air-braking mechanisms of some units.

## Flexible pendant fittings

13.55 These should comply with the requirements of BS EN 737-1:1998 and BS EN 739:1998. In particular, all loose assemblies should be provided with appropriate NIST connectors.

## Bed-head trunking/walling systems

13.56 These fittings should generally be in accordance with BS EN ISO 11197:2004. Separate compartments should be provided for electrical services, nurse call/radio etc and medical gas pipelines.

13.57 Any flexible connecting assemblies used within the fitting should comply with BS EN 739:1998.

13.58 The medical gas compartment should be provided with ventilation by means of louvres, slots etc to prevent the accumulation of any gas in the event of rupture of the medical gas pipeline services.

13.59 In some departments, to engender a more domestic environment, medical gas and other bedhead services are installed within concealed recesses (or behind decorative panels, paintings etc). In such cases, adequate provision must be made for ventilation, and the required space to permit connection and disconnection of equipment should be considered. The covers should be clearly labelled to indicate that medical gas equipment is installed within/behind.

13.60 There are three possible installation procedures:

a. the connection between the pipeline and the trunking should be considered as a second-fix, with the trunking being pre-piped and certificated as having passed a first-fix test; after brazing, the joints will be subject to a second-fix test and leak-tested;

b. the connection between the trunking and the pipework should be as paragraph 13.55;

c. the connection should be by mechanical means, and the separate pipeline connections should be staggered to prevent cross-connection at a later date in the event of the necessity to dismount and reconnect.

### Note

After any such disconnection and reconnection, it will be necessary to carry out the full range of anti-confusion tests.

## LVAs and AVSUs

13.61 All valves should be of the lever-ball type, having flanged O-ring seal connections which open and close with a 90-degree rotation: the handle should be in line with the pipeline when open.

### LVAs

13.62 LVAs should be capable of being locked with the valve in the open or closed position. Means of physically isolating and blanking the pipeline both upstream and downstream of the valve should be provided. The means of isolation should be in the form of a spade that can be readily deployed. It should blank both the pipeline and the valve port and be visible when deployed. Each valve should be provided with a set of "through" and "blanking" spades; they should be coloured white and red respectively. The valve flange should include the thread, and the bolts should be of sufficient length to permit loosening to allow removal/replacement of the spades without loss of structural integrity of the connection. Union-type connections with O-rings are permissible, but the securing nut must also have sufficient thread length to permit venting while maintaining the structural integrity of the connection. A single key for each service is considered to provide sufficient security.

13.63 In the event of leakage of the through or blanking spade, gas must be capable of venting to atmosphere and must not be able to enter either the valve port or the pipeline section blanked.

13.64 The appropriate NIST connector bodies with self-sealing check valves and lockable blanking nuts should be provided upstream and downstream of the flanges. Gas identity and flow direction arrows should be provided for each valve.

### Note

A single NIST connector will suffice in ring-main branch connections between the three closely spaced valves.

13.65 LVAs should be provided as follows:

a. at the connection of the pipeline to any source of supply;

b. at the emergency inlet port (that is, it forms the emergency inlet point);

c. at the pipeline entry to a building;

d. at the pipeline exit from a building;

e. at the connection of branches to the main pipeline run;

f. at the connection to risers;

g. at branches from risers to serve a number of similar departments;

h. upstream, downstream and in the branch connection to a ring-main.

### Notes

a. Where departments on a floor are functionally very different, for example wards and diagnostic areas, it is preferable to run separate branches from the risers.

b. LVAs should not be installed where a leakage of gas could cause an accumulation that is likely to result in a hazardous atmosphere developing. The assembly should include clear labelling of the service.

### AVSUs

13.66 AVSUs are provided for user access in an emergency (or for maintenance purposes). They should be in accordance with the requirements above for LVAs except that security is achieved by installing them in an enclosure with a lockable door designed such that it can be closed with the valve either in the "open" or "closed" position.

### Note

The views of the building operator should be sought as to the level of security that will be required and hence the range of keys.

13.67 In an emergency, the user must be able to gain access in order to operate the isolating valves quickly and simply without the need for a key. There are several methods of providing such emergency access, for example break-glass panels and plastic push/pull-out inserts. Whichever method is used must be safe and secure and must clearly act as a deterrent to tampering without introducing undue risk of injury to the user. Float glass must not be used. The method of emergency access must be obvious and clearly labelled, and its use must be evident.

13.68 AVSUs may be designed for a single pipeline service or multi-services. Where the cover bears the name of the gas service, it should be gas-specific. In the case of multi-service AVSUs, the design should permit the attachment of a hose assembly to any one or more of the NIST connectors and be such that the door can then be closed and locked. The AVSU may include provision for pressure gauges/pressure switches by means of separate bosses.

13.69 The enclosure should have adequate ventilation to prevent the accumulation of gas in the event of a leak. Pipe entries and other penetrations should be sealed to prevent gas escape by routes other than the vents or openings into the user space. The enclosure should be designed to facilitate sealing of these entries on site. Gas identity and flow direction arrows should be provided for each valve.

13.70 Provision and location of AVSUs is covered in Chapter 3.

**Note**

AVSUs should not be installed in positions where they can be obscured or damaged, for example within the "swing" of a department door or behind partitions.

## Specific labelling requirements

13.71 All AVSUs should be labelled to identify the individual rooms, sets of terminal units etc controlled. The upstream and downstream NIST connectors should be clearly identified by a permanent label, securely fixed.

13.72 In critical care areas, where dual circuits and/or subdivisions of circuits occur, terminal units need to be correspondingly identified with the specific AVSUs (see Figures 4 and 5).

13.73 In the case of pneumatically-powered pendant fittings where, typically, medical air (or surgical air) is used for the power source, the AVSU that controls the pneumatically-powered devices should be identified.

## Pressure control equipment

13.74 Medical gases may be distributed at a higher pressure than the eventual nominal pipeline distribution pressure at terminal units. Where this is the case, the maximum pressure should not exceed 1100 kPa, and the local pressure "control" equipment should be installed in an area that has good ventilation and be in a position where it is readily accessible for maintenance/service.

13.75 The pressure control equipment should include duplex pressure-regulating valves, each with upstream and downstream isolating valves, safety valves and NIST connectors. The safety valve discharges should be run to the exterior of the building; medical air and surgical air may be discharged within a plant space, for example a plantroom above an operating department, provided it is terminated in a safe position.

## Pressure sensors

13.76 Pressure sensors to provide the alarm function will need to be fitted to pipeline distribution systems. In all cases they should be installed in a location which is adequately ventilated and having access for maintenance. They may be incorporated downstream of AVSUs. Where not incorporated into an AVSU, the pressure sensor should be close to the AVSU so that it is accessible for maintenance. Pressure sensors should be factory-set and be a replacement item. They should be connected to the pipeline by means of a minimum leak connector.

## Pressure gauges

13.77 Pressure gauges are not usually required outside the plantroom of an MGPS. If provided, however, they should similarly be installed in an adequately ventilated location. They may be incorporated within AVSUs, operating room supply fittings etc. They should be installed with isolation cocks.

## Test points

13.78 Each supply plant, that is, liquid facility, manifold (main and ERM), compressor plant, PSA and blending plant, should be provided with a test point comprising lockable valve and terminal unit. This should be within the plantroom or enclosure, and be sited immediately upstream of the distribution pipeline isolating valve.

## Emergency inlet port

13.79 Medical oxygen and 400 kPa medical air systems should be provided with an emergency inlet port to the pipeline distribution system. This should be located downstream from the main source of supply line valve isolation point to permit connection of a temporary supply plant. The emergency inlet should comprise an LVA, an additional non-return valve on the emergency inlet side and a connection that can be blanked, to which the emergency inlet can be made.

13.80 An emergency inlet port is not required for 700 kPa surgical air systems.

## Line pressure alarms and safety valves

13.81 The purpose of the line pressure alarm is to warn users that the nominal line operating pressure is out of limits and that gas mixtures, whether supplied by a blender/mixer, an anaesthetic machine or patient ventilator, may deviate from the clinically desired proportion. Local action can then be taken to adjust the mixture, or when an anaesthetic machine is in use the reserve cylinders can be brought into use. The low-pressure alarm for nitrous oxide/oxygen mixture supply pipelines will warn of possible demand valve regulator failure so that a portable cylinder can be made available. The high/low pressure limits have been set to accommodate the design of most types of anaesthetic equipment where differential pressure or low pressure may affect performance.

13.82 The line pressure safety valve provides limited safety from differential pressure effects since the pressure at which maximum discharge occurs will result in a differential much greater than that for which the anaesthetic equipment has been designed. They are therefore strictly system protection devices. All safety valves should have a separate discharge pipe that is run to a safe position which – except for air – should be external.

13.83 The commissioning of medical gas pipeline line pressure regulators, warning and alarm systems, and pressure settings is crucial to the performance of anaesthetic equipment and patient safety; once commissioned, medical gas pipelines are subject to strict permit-to-work procedures. Decommissioning a complete system is highly disruptive to patient care and introduces considerable risk.

13.84 The Pressure Systems and Transportable Gas Containers Regulations 1989 require pressure safety devices to be periodically tested. It is not appropriate to test an MGPS by either raising the line pressure regulator setting or manually unseating the relief valve. Such action could result in failure of anaesthetic equipment and – if the safety valve to fails to reseat – considerable gas loss and further hazard. Medical gas pipeline line distribution systems should be provided with a pressure relief device downstream of the line pressure regulator connected by means of a three-way cock so that the safety device can be exchanged for a "certificated" replacement in accordance with the frequency required by the Regulations.

## Other systems

13.85 Quench dump pipes from magnetic resonance imaging (MRI) cryostatic units should be sized to pass the high flows resulting from magnet quenching. They should be routed to safe areas and protected from obstruction. Failure to install adequately sized pipework will result in the venting of asphyxiant helium and/or nitrogen at low temperatures into the working environment, with potentially fatal results.

# 14 Accommodation

## Design and construction of plant and manifold rooms

### Location of manifold rooms

14.1 Cylinder gas/liquid supply systems should not be located in the same room as medical air compressors, PSA systems or vacuum plants.

14.2 Manifold rooms, emergency/reserve manifold rooms for PSA systems, VIE installations and medical compressed air systems should be located to take account of the risk assessment, but should also take account of the location of the medical gas cylinder storage area.

14.3 All manifolds, including the emergency reserve manifolds, may be located within the same room. Manifold rooms should be located on an external wall(s) to facilitate ventilation, which will be required at high and low level. Internally sited manifold room and cylinder stores may require mechanical ventilation.

14.4 The emergency/reserve manifold for liquid oxygen systems has traditionally been located within the VIE compound, but it is preferable to site the manifold separately. For new installations, these emergency/reserve manifolds should be located separately.

14.5 In the case of surgical air the volume of gas used is relatively small even though the instantaneous flow rates are high. Therefore, it may be more convenient to include the manifold within the operating department.

14.6 The surgical air 700 kPa manifold room may be used as the ready-use store for a small number of spare cylinders to be used on anaesthetic machines.

### Note

It is permissible to accommodate medical compressed air plant, vacuum plant and AGS disposal system pumps in general plant areas accommodating such equipment as air handling units, water service systems etc. They should not, however, be located with heating or hot water service equipment or equipment likely to produce any fumes or odour.

### Access

14.7 Access to manifold rooms should be from the open air, not from corridors or other rooms.

14.8 Normal commercial lorry access is suitable for gas cylinder delivery vehicles, but consideration should be given to the provision of a raised level loading bay to reduce cylinder handling hazards.

14.9 Two doors should preferably be provided in a manifold room. One should be large enough to facilitate cylinder handling and must be in an outside wall. Exits must be free of all obstructions. Doors must open outwards. All doors must normally be locked to prevent unauthorised access, but should be provided with means of entry and exit in an emergency (for example by a push-bar arrangement on the inside).

14.10 Internal walls and ceilings, including any internal doors of the manifold room, should be of a suitable non-combustible two-hour fire-resistant material as defined in BS 476-4:1970 and BS 476 Parts 20–23 (1987). Internal doors should be avoided where practicable. Smoke detectors should be provided. Automatic fire suppression should be considered. Manifold rooms should not be located near high-dependency wards.

## Construction and layout of manifold rooms

14.11 The manifold room will contain the manifolds as well as cylinder racks holding sufficient spare cylinders to replace one bank of each manifold and the emergency/reserve manifold. (For nitrous oxide/oxygen manifolds, sufficient spare capacity for two banks of cylinders should be provided.) Further replacement cylinders should be supplied from the medical gas cylinder store. The size of the manifold room should therefore be determined by the risk assessment. Adequate space should also be allowed for cylinder handling.

14.12 A typical automatic manifold with two duty and two stand-by cylinders is 1.8 m long and 0.6 m deep. One extra cylinder on each bank adds approximately 0.5 m to the overall length, so that a 2 x 6 manifold is approximately 4 m long.

14.13 All medical gas manifolds may be installed in the same room. Additional floor area should be provided to accommodate separate storage racks for each gas. The racks should be designed along the lines of those on the manifolds, but the stored cylinders may be closer together. Racks should conform to ISO 32:1977. With the exception of small cylinders of $N_2O/O_2$ mixtures, under no circumstances should rooms contain gas cylinders other than those appropriate to their manifolds.

## Heating and ventilation

14.14 Ventilation louvres should be provided at both high and low level for all manifold rooms, to allow circulation of air. As a guide, separated openings equivalent to at least 1.5% of the total area of the walls and floor should be provided. For example, given a manifold room 5.0 x 4.0 x 2.4 m with a total area of the walls and ceiling of 63.2 $m^2$, the total free open area for ventilation required is 1 $m^2$.

14.15 Air intakes for compressor inlets should, if possible, be located externally. However, they should not be installed as an alternative to the provision of adequate ventilation for cooling purposes.

14.16 All ventilation louvres should be vermin/bird-proof.

14.17 PSA and medical air compressors liberate, under maximum flow conditions, considerable heat. Moreover, these plants aspirate air for breathing purposes. Generous natural ventilation should be provided, and where this is not possible if the plantroom is deep-plan, mechanical ventilation should be provided. The ambient temperature of manifold rooms and plantrooms should be maintained within the range of 10–40°C. The ventilation rates should ensure that the plantroom temperature does not exceed ambient temperature by more than 10°C.

14.18 Manifold rooms may be used to store small numbers of nitrous oxide/oxygen cylinders intended for portable use; these are taken from the main cylinder store for the purpose of temperature equilibration, before being delivered to wards etc.

14.19 To achieve temperature equilibration, additional heating may be required; the natural ventilation must not be reduced. Where such heating is provided, it should be by indirect means, for example steam, hot water or warm air. Naked flames and exposed electric elements should not be used, and excessive surface temperature should be avoided. If necessary, cylinders should be protected from excessive heat. Any primary heat source should be located in a safe position, preferably remote from the manifold room.

14.20 Cylinder recognition charts, supplied by the medical gas supplier, should be prominently displayed as appropriate.

## Lighting

14.21 Manifold rooms should be provided with lighting to an illumination level of 150 lux by means of bulkhead lighting fittings to BS EN 60529:1992. Plantrooms other than manifold rooms should be provided with a lighting level of 200 lux.

## Noise control

14.22 Plantrooms should be designed and constructed to ensure the satisfactory control of noise emission. The effect of two vacuum pumps or compressors running together, in the case of duplex installations, and three or more in the case of multiplex installations, will be to increase the free-field noise level outside the plantroom by 5 dBA for each additional pump or compressor operation over and above the specified limits. Consideration should be given to providing acoustic enclosures to reduce the free-field noise levels in noise-sensitive areas adjacent to plantrooms.

14.23 Acoustic enclosure and/or plantroom design must not inhibit normal cooling functions or maintenance activities.

14.24 Free-field noise levels should be given to the architect to assist in acoustic design of the plantrooms.

14.25 The discharge from some vacuum pumps may require silencing.

14.26 Compressors and pumps should be mounted on properly-selected anti-vibration mounting, where necessary, to minimise transfer of noise and vibration to the structure of the building.

14.27 All pipework and electrical conduits connected to the plant should be fitted with flexible connectors where necessary to prevent the transmission of noise and vibration along the pipelines and conduits. Electrical bonding will be required.

## Labelling/signage

14.28 Labelling for medical gas systems, equipment and accommodation should be in accordance with Appendix K.

# 15 Validation and verification

15.1 This chapter covers the validation and verification and filling for use of MGPS. The requirements for AGS disposal systems are also covered in paragraphs 15.169–15.178.

15.2 This chapter describes the tests required and the test methods. The contractor should provide instrumentation for the functional tests. The Quality Control Pharmacist normally provides instrumentation for the quality tests. Calibration certificates should be available for all instrumentation. Tests are listed in Appendix A with the associated forms.

15.3 The objective of testing and commissioning is to ensure that all the necessary safety and performance requirements of the MGPS will be met. Testing and commissioning procedures will be required for new installations, additions to existing installations and modifications to existing installations. The scope of work will dictate the specific test programme required. This is described in more detail in paragraphs 15.12–15.14.

15.4 For modifications and extensions (except for the final connection), all work should be performed with an inert gas shield; thus, it is essential that a physical break is employed between the pipeline being modified/extended and the system in use. (This will usually be by deploying "spades" in AVSUs and LVAs, or by cutting and capping the pipe.) Prohibition labels should be affixed to all terminal units in the system affected in occupied areas.

15.5 For small extensions comprising fewer than 20 brazed joints per gas service, all the tests may be performed with the working gas – the carcass pressure tests being replaced by a system leakage test of the complete extension. An extension comprising more connections would, however, be deemed to be a small installation, requiring all the appropriate tests to be carried out, up to the final connection; the final connection would be tested at pipeline distribution pressure. For the purpose of ascertaining the number of joints, a straight coupling comprises two joints and a "T" comprises three joints. On a minor modification, from which existing terminal units would not be removed, the carcass pressure test can also be omitted. All other tests would be required, including a working pressure test.

15.6 The programme of tests is divided into three phases:

a. tests and checks on the pipeline carcass;

b. tests and commissioning of the complete pipeline system (with terminal units installed) for safety, performance and particulate contamination using test gas;

c. filling of the systems with specific gases for the purposes of identity and quality tests of the specific gases prior to use for patient care.

15.7 The basic rationale for the tests is depicted as a decision tree in Figure 36.

## Notes

Systems that are not to be taken immediately into use should be filled with medical air and maintained at operating pressure. Systems other than medical, surgical or dental air supplied from compressors should be filled with medical air from cylinders.

Commissioning of liquid supply systems prior to handover should be avoided. (Under "no flow" conditions, liquid will evaporate and oxygen will blow off to atmosphere.)

15.8 The personnel and test equipment needed for these tests are listed together with the test requirements in Table 27. The particulate contamination test for all pipeline systems may be checked using medical air to establish that the pipeline has been constructed correctly and is not contaminated. Successful completion of the commissioning tests normally indicates the end of the installation contract. The systems may then be left under pressure, filled with medical air, for an indefinite period. Responsibility for the system during this

## Figure 36 Decision tree for testing and commissioning

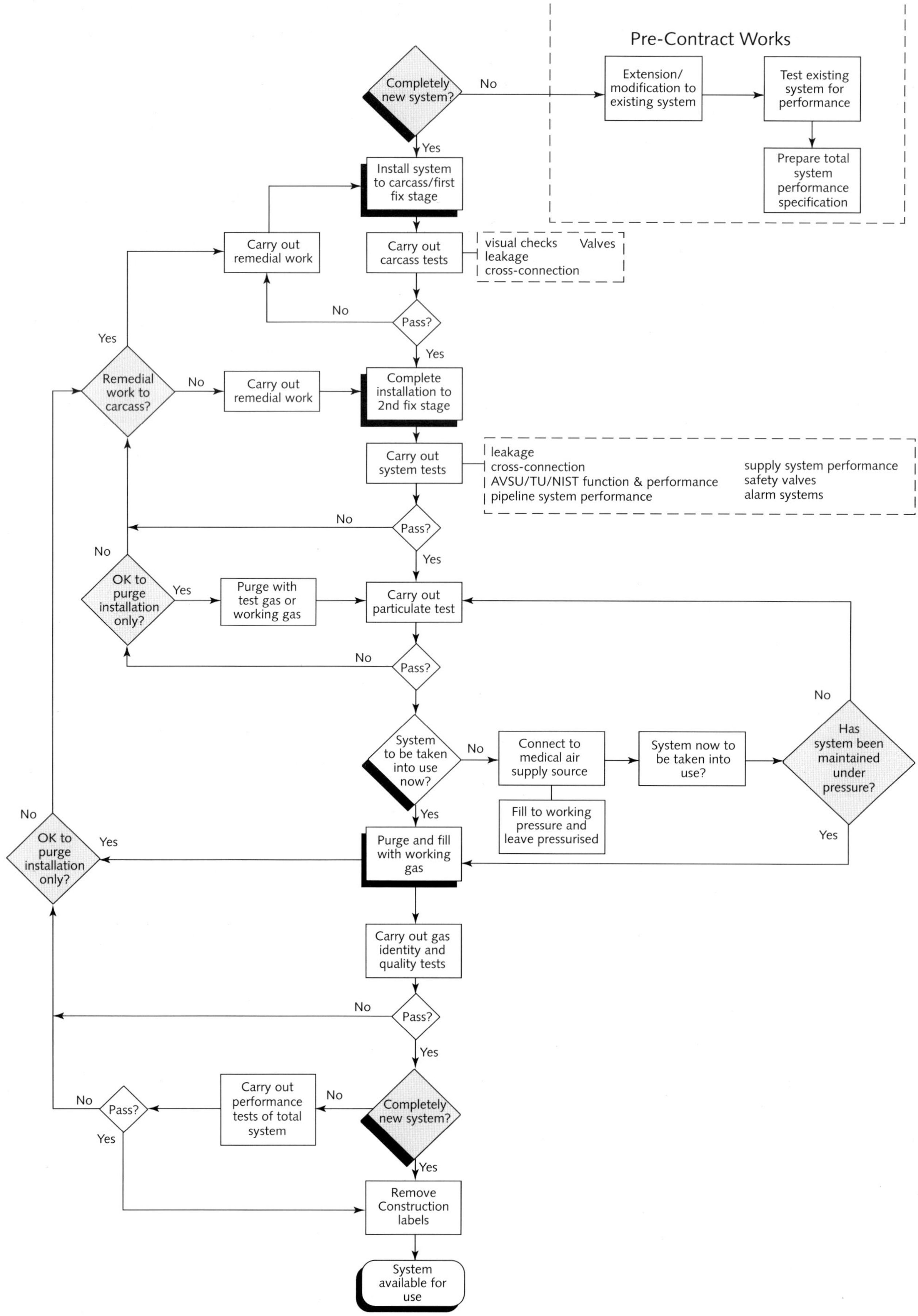

period needs to be clearly defined in the contract; the Authorised Person (MGPS) ultimately responsible for the day-to-day management of the MGPS after handover should be permitted access during contract work. This should be included in the contract agreement.

### Table 27 Personnel and test equipment requirements

| Paragraph | Test | Personnel | Equipment |
|---|---|---|---|
| **15.12: Pipeline carcass** | Labelling and marking | CSO and CR | Visual |
| | Sleeving and supports | CSO and CR | Visual and tape |
| | Leakage | CSO and CR | Pressure-measuring device |
| | Cross-connection | CSO and CR | Pressure-measuring device |
| | | | |
| **15.13: Pipeline system** | Leakage | CSO, CR and AP | Pressure-measuring device |
| | Closure of AVSUs and LVAs | CSO, CR and AP | Pressure-measuring device |
| | Zoning of AVSUs and terminal unit identification | CSO, CR and AP | Open probes or special test device |
| | Cross-connection | CSO, CR and AP | Open probes or special test device |
| | Flow and pressure drop at individual terminal units, mechanical function and correct installation | CSO, CR and AP | Special test device, certified probes |
| | NIST connectors | CSO, CR and AP | Full bore NIST probes and nut |
| | System performance | CSO, CR and AP | Metered leaks and special test device |
| | Supply systems | CSO, CR and AP | Visual |
| | Pressure safety valves | CSO, CR and AP | Visual |
| | Warning and alarm systems | CSO, CR and AP | Visual |
| | As-fitted drawings | CSO and AP | Visual |
| | Purging and filling | CSO and CR | Gas source and delivery equipment |
| | Particulate contamination (see note after paragraph 15.8) | CSO, CR, AP and QC (MGPS) | Particulate matter test device (PMTD) |
| | Gas quality | CR, QC (MGPS) and AP | PMTD, oil, moisture, CO, $CO_2$, $SO_2$ and $N_2$ oxides measuring devices, $O_2$ and $N_2O$ analysers |
| | Gas identification | CR, QC (MGPS) and AP | $O_2$ analyser and $N_2O$ meter |
| | AGS disposal systems | CR and AP(1) | Metered leaks and AGS test device |

**Key:**

| | |
|---|---|
| CR | Contractor's representative. |
| AP | Authorised Person (MGPS). |
| CSO | Contract Supervising Officer (for tests on the pipeline distribution system, CSO responsibility would normally involve witnessing of the test documentation). |
| QC (MGPS) | Quality Controller (Medical Gas Pipeline Systems) |

**Note:**

1. The QC (MGPS) can perform this test

**Note**

In some circumstances an MGPS may not be taken into use immediately after construction and will be left filled with medical air. In these circumstances, the particulate contamination and odour/taste tests may be carried out before purging and filling with the working gas (see paragraphs 15.81–15.82).

15.9 All supply systems and their major components should have certificates (as specified in Model Engineering Specification C11 – 'Medical gases') which show that they meet the design requirements of the pipeline system.

15.10 Only contractors who are registered to BS EN ISO 9001:2000/BS EN ISO 13485:2003 with their scope of registration defined to include commissioning should undertake engineering validation and verification.

**Note**

BS EN ISO 9001:2000 registration is also recommended for independent medical gas testing agencies but is not necessary for appropriately trained and appointed hospital-based QCs (MGPS). (A national register of QCs (MGPS) is to be prepared during the lifetime of this Health Technical Memorandum.)

15.11 All relevant tests should be carried out by the persons listed in Table 27 and witnessed by the appropriate persons, who must record the results of the tests in writing for the hospital authority.

## Summary of tests

### Tests and checks on the pipeline carcass

15.12 The following tests must be carried out after installation of the pipeline carcass but before concealment:

a. visual check of pipeline labelling, marking, sleeving and support;

b. leakage test;

c. tests for cross-connection;

d. valve tests for closure, zoning and leakage. (These tests will be repeated as part of the pipeline system tests and the contractor may wish to defer closure and leakage, but may choose to carry out a zoning check.)

### Tests on the pipeline system

15.13 The following tests and checks must be carried out after complete installation of the pipeline system:

a. tests for leakage on each MGPS;

b. tests of AVSUs for closure, correct service and control of the terminal units in the zone: checks for correct labelling of AVSUs for zone reference and identity of terminal units controlled and flow direction indication;

c. tests of LVAs for closure and identification;

d. tests for cross-connection, flow, pressure drop, mechanical function and correct identity of the terminal units: checks for correct labelling and association with AVSUs (this is only required when, within a specific area, there are separate circuits for the same service, for example dual/split circuits);

e. tests for mechanical function and identity of NIST connectors;

f. performance tests of the pipeline system;

g. functional tests of all supply systems;

h. checks of safety valve certification;

j. tests of warning systems;

k. tests for particulate contamination/odour/taste: these may be carried out immediately after installation, using medical air, or after purging and filling with the specified gases. It is intended that the Quality Controller (MGPS) is handed over a system that is purged clear of gross particulate contamination before being filled with the working gases. However, it is accepted that this may not always be possible. If the system is not to be taken into immediate use, the tests for particulate contamination and odour/taste should be carried out with medical air, and the system then left under pressure (see paragraph 15.93);

**Note**

Nitrous oxide and nitrous oxide/oxygen mixture are not tested for odour.

m. tests for anaesthetic gas scavenging disposal systems.

### Tests before use

**15.14** The following tests must be carried out after purging and filling with the working gas:

a. tests for particulate contamination (see paragraphs 15.93 and 15.130–15.136);

b. tests for gas identity;

c. tests for gas quality.

## General requirements for testing

**15.15** The tests described in this document are generally carried out, in the order given, for new installations. It may be necessary to amend the test programme for modifications or extensions to existing systems. Care must be taken, however, to ensure that the basic principles are followed. Paragraphs 15.32–15.44 give details of the tests required for modifications/extensions to existing systems.

> **Note**
>
> When tests and/or purging are/is carried out on systems fed by sources serving an operational hospital, it is essential to ensure that the flows generated during any tests do not result in interruption of continuity or impairment of adequacy of supply in the operational areas.

**15.16** Testing for leakage is normally carried out in two stages: the first to the pipeline carcass, the second to the completed distribution system, which will include terminal units and medical supply units as appropriate.

**15.17** Purging and testing must be carried out with clean, oil-free, dry air or nitrogen, except for those tests where medical air or the specific working gas is prescribed. All test gases must meet the particulate contamination requirements set out in paragraphs 15.130–15.136. The shield gas may be used for the tests on the pipeline carcass described in paragraphs 15.49–15.56. Cylinders of medical air will normally be used as the source of test gas for oxygen, nitrous oxide, nitrous oxide/oxygen and helium/oxygen systems in order to prevent the possibility of contamination with oil.

**15.18** However, in the case of a large oxygen system, for example a new-build, the use of cylinders will be impracticable for the total system performance test. As it may be undesirable to commission the liquid supply system, the total system performance test can be carried out by using the medical air compressor system, provided that the quality tests have been satisfactorily carried out to demonstrate that the criteria set out in Table 30 have been met and that the air supply plant is continuously monitored for moisture during the test.

**15.19** Such a compressor system can also be used for the single point performance tests etc and for initial purging and particulate testing of these systems. Once tests have been completed, the system should be maintained under pressure by means of air supplied from medical gas cylinders until filled with the working gas, when full QC checks will be carried out.

> **Note**
>
> The use of portable, non-medical air compressors is not appropriate. Not only should a Quality Controller (MGPS) check all compressors **before** use, but also QC checks **during** use are important. Preferably, an on-line dew-point meter should be fitted to the plant or pipeline system.

**15.20** When employing compressors for this type of test, it is important that the system demand should not exceed the maximum flow capacity of the dryers, otherwise wet air will result. It is suggested that the total flow required by the system under test should not be more than 75% of the flow capacity of the dryers.

**15.21** It is also important not to introduce such a compressor after identity checks have taken place.

**15.22** Special care will be required when carrying out QC checks, as some synthetic oils cannot be detected using portable equipment.

**15.23** Special connectors will be needed to introduce test gas into different pipeline systems. These must be of distinctive construction and permanently labelled with their function and the contractor's name. The location of special connectors on the site must be recorded and should be subject to routine inspection under a planned preventative maintenance (PPM) system. They should be removed from site when work is complete and the contractor should record their removal.

**15.24** New terminal units are supplied with "Do not use" labels. These labels should remain in place until the final identity and quality tests have been completed. They are then removed by the Authorised Person (MGPS).

15.25 In the case of existing systems, "Do not use" labels should be affixed to all terminal units within the section being modified.

15.26 The results of all tests must form part of the permanent records of the hospital and should show details of the services and areas tested. Examples of the appropriate forms are given in Appendix A. All signatories are entitled to copies of the test forms. The procedure for filing and retaining these forms should be included in the local MGPS operational policy.

15.27 For total system pressure tests on oxygen, nitrous oxide and nitrous oxide/oxygen mixture, the system under test must be physically isolated from the source of supply (for example by the use of spades). In the case of compressed air and vacuum systems, the pressure at the plant must be respectively below and above pipeline distribution pressure.

15.28 All errors found during testing must be rectified, and the relevant systems must be retested as appropriate before the records are signed.

15.29 The contractor (MGPS) must provide all engineering forms, labour, materials, instruments and equipment required to carry out the tests described in this chapter. In the case of engineering tests, this must include all cylinders of test gas together with "open" bore NIST connector probes, pressure-measuring equipment and gas specificity/flow pressure testing device(s), metered leaks and AGS disposal system test equipment. The Quality Controller (MGPS) will be responsible for supplying all QC forms, unless otherwise requested by the trust, calibrated test equipment, connections etc.

## Note

If there is to be a delay between completion of the MGPS and when it is taken into use, it will be necessary to carry out the particulate and odour test prior to purging and filling with specific gases. In such cases the contractor must also provide labour, materials and equipment to carry out these tests.

15.30 The Quality Controller (MGPS) should provide the test equipment specified in Appendices D, E and F. The Quality Controller (MGPS) should provide all equipment for gas quality and identity testing. It should be regularly serviced and calibrated to an appropriate standard and the Quality Controller (MGPS) should maintain calibration records. On-site pre- and post-testing calibration of equipment against an appropriate standard will be performed at the discretion of the Quality Controller (MGPS).

15.31 In a completely new installation, flow meters, anaesthetic trolleys etc should not be moved into rooms until validation and verification tests have been satisfactorily completed.

## Note

In existing installations, particular care must be taken to ensure that medical gas equipment left in areas where work or testing is taking place is, and remains, disconnected from the system.

Medical and nursing staff should be made aware of this situation by the posting of appropriate exclusion notices and terminal unit "Do not use" labels.

## Modifications, extensions or repairs to existing systems

15.32 Where modifications, extensions or repairs to existing systems are carried out, the tests and the sequence of tests summarised in paragraphs 15.12–15.14 should be followed as far as possible.

15.33 The permit-to-work system should always be followed whenever any work is carried out on an existing system. The Authorised Person (MGPS) should act on behalf of the management and therefore would not normally be a member of the installation contractor's staff.

15.34 Whenever modifications or extensions are carried out, it may be advisable but not always possible to test both the existing system and the new system separately before the break-in is made. Existing systems should, if possible, be tested to determine their performance and to identify any potential limitations that may arise as a result of modifications.

15.35 Where there is any doubt as to the cleanliness, it is in the interest of both the contractor and management for particulate tests to be carried out on the existing system prior to any break-in, and it is the responsibility of the hospital authority to ensure that these tests are carried out prior to the design phase of any modifications or extensions.

15.36 It is the responsibility of the hospital's management to ensure that any required remedial work is carried out on an existing system before extensions are added.

15.37 No system should be modified during the process of testing. It is important that any modifications are documented and that any additional testing required, as a consequence of those modifications, is performed.

15.38 A permit-to-work (or another form of appropriate document) must be issued if additional works are to be carried out during the commissioning process, even though a permit will not have been issued for the original commissioning.

15.39 The tests for particulate contamination of any extension or modification may be carried out with medical air, prior to connection and handover to the Quality Controller (MGPS), although in extensions comprising fewer than 20 joints, the working gas will generally be used to perform all tests.

15.40 The Quality Controller (MGPS) will normally carry out all checks, including a repeat of the particulate matter test, using the working gas.

15.41 The exact tests to be carried out will depend on the nature of the modification/extension. A specification should be prepared for the performance of the completed system. This specification should be as close as possible to that given in Table 28.

15.42 Some older compressed air systems will have been designed to provide 250 L/min at the terminal unit in accordance with Health Technical Memorandum 22 (1978). It may not be possible for such systems to provide 350 L/min, as specified in Table 19, and there may be circumstances where this would be acceptable. This should be clearly stated in the specification for the performance of the completed system. However, every effort should be made to comply with the performance and quality specifications given here, although particular care must be taken to avoid degradation of air quality arising from dryer units working at flow rates above their design specification.

15.43 It may be necessary to repeat some of the system performance tests (such as flow and pressure drop) at selected terminal units on the completed system to demonstrate satisfactory performance (see paragraph 15.77). To ensure a valid result from such a test, it should be performed when flows in the system are representative of typical maximum demands.

15.44 The break-in to the existing system should be carried out with an inert gas shield where possible, for example where AVSUs have been installed, and a downstream blanking plate has been deployed. A leak test must be carried out using a suitable leak detection fluid on this final joint at working pressure, and the joint purged with the working gas.

15.45 Connection of the upstream side of the AVSU into the existing system will usually be made without use of the shield gas. This joint can be purged with the working gas (exiting via the AVSU upstream NIST).

## Note

In some articulated pendant fittings, it is not always possible to achieve the specified pressure requirements for surgical air and vacuum. In the case of surgical air, it is most likely to be a potential problem in orthopaedic operating rooms. As these normally include an ultra-clean system into which can be incorporated surgical air (and other terminal units), supplied by rigid pipework, there may not be a problem in practice. If the static pressure exceeds the nominal pressure during flow by more than 25%, the possibility of installing hoses with a greater bore should be considered. In the case of vacuum, the flows required during surgery are less than those used during testing.

## Requirements for pipeline carcass tests

15.46 If sectional testing is performed, it is essential that as-fitted drawings are available so that the extent of the system(s) under test can be identified. For the purpose of the leakage test, all pressure gas systems may be interlinked, provided that the test can be performed at the highest pressure required. (This also has the advantage that the pipeline carcass could be assigned to a different service.)

## Notes

In the event of a leak, it will be necessary to test each system separately.

It is advantageous to perform the tests with nitrogen, since – in the event of a leak or cross-connection – remedial action can be taken immediately.

When connecting systems together, vacuum systems should not be included, as particulates from an unpurged vacuum system may be drawn into any part of any pressure gas system by venturi effects.

**Table 28 Validation and verification: pressure during pipeline system tests**

| Medical gas | Nominal pipeline distribution pressure (kPa) | Test flow (L/min) (measured at terminal unit outlet) | Test location | Min. pressure at design flow measured on the test gauge (kPa) | Plant pressure (kPa) |
|---|---|---|---|---|---|
| $O_2$ | 400 | 100 | Operating rooms (ORs) and anaesthetic rooms (ARs) | 380 | 430–490 |
| | | 10 | All other terminal units | | |
| $N_2O$ | 400 | 15 | All terminal units | 380 | 430–490 |
| Medical air (400 kPa) | 400 | 80 | Critical care areas, SCBU, HDU, ORs and ARs | 380 | 430–490 |
| | | 20 | All other terminal units | | |
| Surgical air (700 kPa) | 700 | 350 | All terminal units | 700 kPa at 350 L/min (max 900 kPa at no flow conditions)(1) | See Chapter 4 |
| $O_2/N_2O$ mixtures | 400 | 275 | LRDP rooms (inhalationary gasps) | 310(2) | 430–490 |
| | | 20 | All other terminal units | 380 | |
| $He/O_2$ | 400 | 80 | Critical care areas only | 380 | 430–490 |
| Vacuum | 55.3 kPa (400 mm Hg) below standard atmospheric pressure of 101.3 kPa (760 mm Hg) | 40 | All terminal units | 400 kPa (300 mm Hg) | 550–650 mm Hg (typical plant operating range) |

**Notes:**

1. Downstream of the local regulator measured at the terminal unit.

2. The peak flow of 275 L/min must be achieved for 5 seconds at a minimum pressure of 310 kPa (310 kPa is the minimum pressure required for the operation of a demand regulator).

## Labelling and marking

15.47 A visual check must be made on each pipeline system to ensure that the pipelines are labelled in accordance with the contract specification, and that the terminal unit base blocks are marked in accordance with BS EN 737-1:1998. The results of the checks are recorded on Form B1.

## Sleeving and supports

15.48 A visual check must be made on each pipeline system to ensure that the pipelines are sleeved, where required, and supported in accordance with Table 25. The results of the checks are recorded on Form B1.

## Leakage

15.49 The aim of this test is to establish that there is no leakage from the piped medical gas systems. This is demonstrated by the use of electronic pressure-measuring equipment with a minimum resolution of 0.2 kPa in 1000 kPa and 0.5 kPa in 2000 kPa.

### Note

With suitable equipment it is possible to carry out this test during a relatively short period to minimise the effect of temperature change.

15.50 During a test period of two hours, the maximum pressure loss should be less than 0.2 kPa for 400 kPa systems and vacuum, and 0.5 kPa for 700 kPa systems. No allowance is normally made for variation of pressure with temperature; if, however, the accuracy of the available pressure-measuring equipment is in doubt and recourse is made to a 24-hour test, Appendix B contains information on the method of calculation. Systems must be tested at a working pressure of 18.0 bar for medical compressed air systems for surgical use, 10.0 bar for all other compressed medical gas systems and 5.0 bar for vacuum systems constructed in copper (1 bar for systems constructed in plastic).

15.51 This test should be carried out with AVSUs, LVAs and other service valves open; any safety valves and pressure-sensing devices installed may be removed and the connections blanked off. The results of the test may be recorded on Form B1.

## Cross-connection

15.52 Before performing these tests, any links between systems should be removed and all pipelines should be at atmospheric pressure with all AVSUs etc open.

15.53 A single pressure source should be applied to the inlet of the system to be tested and at least one terminal unit base block on all other systems should be fully open.

15.54 Each terminal unit base block on the pipeline under test should be opened in turn, checked for flow and then re-blanked. (To permit refitting of blanking caps, it is necessary to partially open at least one base unit – but it is still necessary to achieve a detectable flow.) When the test on one pipeline has been completed, the pressure source should be removed and the pipeline should be left open to atmospheric pressure by removing at least one base block blanking plate.

15.55 The test is repeated for other systems, one at a time.

15.56 The results may be recorded on Form B2.

## Requirements for pipeline system tests

15.57 There must be no links between the pipeline systems. Engineering (pressure) tests should be carried out with electronic pressure-measuring equipment with a minimum resolution of 0.2 kPa in 1000 kPa, and 0.5 kPa in 2000 kPa.

15.58 The scope of the system and scale of provision of terminal units, AVSUs, LVAs and warning and alarm system panel indicators should be checked for compliance with Table 11 and any deficiencies noted.

## Leakage from total compressed medical gas systems

15.59 This test must be carried out on the completed system with all terminal units, AVSUs, pressure safety valves and pressure transducers fitted. Once the test pressure has been applied, the system should be isolated from the plant. For the purpose of this test, the supply system extends from the last valve(s) nearest to the plant detailed on the appropriate schematic drawing. This point should be identified on the contract drawings. The test is performed at pipeline distribution pressure.

15.60 During a test period of two hours, the maximum pressure loss should be less than 0.2 kPa for 400 kPa systems and vacuum, and 0.5 kPa for 700 kPa systems. The test results may be recorded on Form B3.

## Leakage into total vacuum systems

15.61 Prior to testing, the vacuum plant should be operated to allow any moisture in the system to evaporate. With the system at pipeline distribution pressure and with the source isolated, the pressure increase in the pipeline must not exceed 1 kPa after one hour. There is no additional allowance for temperature correction in this test.

15.62 The test results may be recorded on Form B4.

## Closure of area valve service units and line valve assemblies

15.63 For pressurised systems, the system upstream of the closed AVSU under test must be maintained at pipeline distribution pressure and the downstream line pressure should be reduced to about 100 kPa. The downstream pressure must be recorded, and there should be no change in pressure over a period of 15 minutes.

15.64 For vacuum systems, the systems on the supply plant side of the closed valve must be maintained at pipeline distribution pressure and the terminal unit side should be at about 15 kPa. The upstream (terminal unit side) pressure must be recorded, and there should be no change in vacuum over a period of 15 minutes.

15.65 For LVAs, a similar test procedure is adopted. There is no change in the time for vacuum.

### Note

The reduced residual pressure is intended to take into account any potential terminal unit leakage on the assumption that it is unlikely any such leakage would equate to that of the valve under test; there would be less certainty if the pressures were reduced to zero.

15.66 The test results may be recorded on Forms B5A and B5B.

## Zoning of AVSUs and terminal unit identification

15.67 This test is performed to ensure that each AVSU in the pipeline controls only those terminal units intended by the design. Each terminal unit must be checked to ensure that it is for the correct service and that it is in accordance with BS EN 737-1:1998; unambiguous cross-referenced labelling of AVSUs and terminal units controlled by them is essential. It is particularly important to establish correct identification where dual or separate circuits have been installed; often it is not obvious by the spatial relationship of AVSUs and terminal units which of the AVSUs controls which terminal unit arrays.

### Notes

The contractor may wish to carry out this test as part of the carcass tests before any section of the pipeline is "enclosed".

Terminal-unit first-fix back blocks inadvertently fitted upside-down will result in inverted second-fix components, unless gas-specific components are deliberately removed. Therefore, a selection of terminal unit second-fixes, for example one per ward area, should be removed and examined to ensure that no gas-specific components have been removed.

15.68 The test is performed by turning off an individual AVSU and venting the zone to atmospheric pressure. A check is then made to establish that only those terminal units controlled by the AVSU are at atmospheric pressure. All other terminal units, including those for other gas services, should be at the operating pressure. Once a zone has been vented, it is not necessary to repressurise. The other AVSUs are then tested successively.

### Notes

These tests can be performed at the same time as the cross-connection/terminal unit pressure drop tests.

Where pneumatically activated pendant fittings are installed, a check should be made to ensure that the source of air has been taken from the correct AVSU zone.

15.69 The test results may be recorded on Forms B5A and B5B.

## Cross-connection

15.70 All systems must be checked to ensure that there is no cross-connection between pipelines for different gases and vacuum. The tests should not commence until all installations are complete and plant operational. (The tests can be performed using "test" gas or "working" gas.)

### Note

Oxygen and vacuum can be tested simultaneously, followed by medical air and surgical air simultaneously, followed by the other gases, that is, nitrous oxide and nitrous oxide/oxygen mixture. (Helium/oxygen mixture usage is increasing, and pipeline systems may be encountered. Also, carbon dioxide pipelines are being installed.)

15.71 The sequence of the test is, first, to open all valves on all systems (for example AVSUs, LVAs and any other valves). For oxygen and vacuum systems, the main plant isolation valves should be opened (the main plant isolation valves on other systems remain closed). A check must be made to ensure that there is a flow at every oxygen terminal unit and suction at every vacuum terminal unit, and that the systems are at the correct operating pressure; there must be no flow at any other terminal unit for the other gases.

15.72 For the next stage, the main isolation valves for medical air and surgical air, if present, are opened. (It is not necessary to return the oxygen and vacuum systems to atmospheric pressure.) A check is made to ensure that there is a flow at every medical air terminal unit and every surgical air terminal unit and that the operating pressure is correct; there must be no flow from the nitrous oxide and/or nitrous oxide/oxygen mixture terminal units, if present, and helium/oxygen, if present.

15.73 The process is then repeated for nitrous oxide – again there is no necessity to return any of the previously tested systems to atmospheric pressure. A check is made to ensure that there is flow at every nitrous oxide terminal unit and that the operating pressure is correct; there must be no flow from the nitrous oxide/oxygen terminal units and helium/oxygen terminal units (if present).

15.74 The process is then repeated for nitrous oxide/oxygen mixture, and finally helium/oxygen mixture. (If other medical gases are encountered, for example carbon dioxide, the sequential testing methodology will continue.) As before, there is no necessity to return any of the previously tested systems to atmospheric pressure. Checks are made to ensure that there is no flow from any system that is still isolated at the plant.

### Note

The tests can be carried out on a total system basis, departmental basis or sub-departmental basis, having previously checked for cross-connection up to the appropriate AVSUs. When carrying out the tests on a sectional basis, it is essential that as-fitted drawings are available such that the extent of the system(s) can be established.

15.75 This test must be repeated in full if any subsequent modifications are made to the pipeline system.

15.76 The test results may be recorded on Form B6.

## Flow and pressure drop at individual terminal units, mechanical function and correct installation

15.77 These tests can be carried out as part of the cross-connection tests above using appropriate test devices as described in Appendix C with the correct probes inserted for the pipeline(s) under test. The pressure must achieve the values given in Table 28 at the specified flows.

### Note

When performing these tests as part of the cross-connection tests, there is the possibility that the 400 kPa and vacuum test devices could be connected to the incorrect service, particularly a vacuum and oxygen reversal. The instruments used, therefore, should include appropriate directional check valves. (There is a possibility of damaging the gauges. Alternatively an open probe can be used to determine pressure or vacuum.)

15.78 It must be demonstrated for each terminal unit that the appropriate gas-specific probe can be inserted, captured and released, and it should be visually confirmed that an anti-swivel pin is present, or absent, in terminal units with a horizontal or vertical axis, respectively.

### Notes

Terminal units to BS EN 737-1:1998 need not be challenged with the full complement of BS 5682:1998 test probes.

The terminal unit should be fitted complete with bezel plates etc. The clearance hole should be reasonably concentric with the terminal unit rim – it must not be in contact.

15.79 The results of the tests may be recorded on Form B7A.

15.80 All NIST connectors must be checked to ensure that gas flow is achieved when the correct NIST probe is inserted and mechanical connection made. The correct identification of gas flow direction should be confirmed for AVSUs (that is, which are the upstream and downstream NIST connectors). NIST connectors can be checked when performing other tests on AVSUs and LVAs.

### Note

Whereas it should not be necessary to carry out these tests on AVSUs bearing a CE Mark, in certain circumstances factory-assembled units are dismantled for installation purposes and can be subsequently incorrectly re-assembled. In the case of LVAs (whether or not CE marked), disassembly and subsequent incorrect re-assembly or, indeed, insertion into an incorrect line, is also possible.

The primary purpose of the test is to ensure that whenever it is necessary to make a connection, the appropriate connectors will be to hand; the test is a further safety aid, although it is assumed that personnel making connections to NIST fittings are appropriately qualified and authorised to do so.

15.81 It must be demonstrated (except for vacuum) for each NIST connector that the self-sealing device substantially reduces the flow of gas when the connector is removed without hazard to personnel or reduction in pipeline pressure.

### Note

Personnel should take care not to stand in front of the NIST connector when performing this test.

15.82 The results may be recorded on Form B7B.

## Performance tests on the pipeline system

15.83 The performance of individual pipeline systems is measured by introducing a sufficient number of calibrated metered leaks (with orifice sizes providing different flows that replicate the range of medical devices for which the pipeline is designed; see Table 12) to represent the total "diversified" system design flow, less the flow generated by the test device. Thereafter, a representative number of terminal units (see note below) are tested for pressure and flow: the diversified flows should be derived from the data in Tables 13, 15, 16, 18, 20 and 21.

### Notes

In a 28-bed ward module a representative number would be in the order of two terminal units furthest from the AVSU, two near the entrance, and the treatment room, if applicable for each gas and vacuum.

In an operating department, a representative number would be one terminal unit in each operating suite and 20% of terminal units in recovery for each gas and vacuum. For oxygen, one metered leak should be 100 L/min to represent oxygen "flush".

It is not necessary to insert metered leaks into the actual number of terminal units used to calculate the "diversified" design flow, provided the numbers used are evenly distributed and orifice sizes are selected to achieve this flow.

15.84 The metered leaks should be stamped or similarly be identified to show the flow (air equivalent) at, for example, 10, 20, 100, and 275 L/min for 400 kPa systems, and 350 L/min for 700 kPa systems; the results of the tests may be recorded on Form B8.

### Note

In principle it is permissible, although unlikely to be practicable for large installations, to test all systems simultaneously, particularly oxygen and vacuum, where terminal units are installed in pairs and where they require different metered leaks (this includes vacuum when testing oxygen will not significantly increase the time needed).

## Functional tests of supply systems

15.85 All supply systems must be tested for normal and emergency operation, according to the manufacturers' manuals and contract specifications. For the purpose of the tests, the systems must be connected to both the normal and stand-by power supplies. The results of these tests should be recorded on Form B9.

## Pressure safety valves

15.86 Pressure safety valves are not tested. They should be examined to ensure that they are correctly rated for the pipeline system and are in accordance with the contract specification. Each should be provided with a test certificate confirming the certificated discharge pressure. Records of safety valve details should be noted on Form B10.

15.87 Check that the specified pressure safety valves, line valves and non-return valves have been fitted.

15.88 Verify that the pressure safety valves are certified to operate in accordance with the contract specification and conform to BS EN ISO 4126-1: 2004.

## Warning and alarm systems

15.89 The operation of warning and alarm systems should be tested in all normal operating and emergency modes. Particular attention should be paid to the following:

a. that all systems operate within the specified tolerance limits at all operating parameters and fault conditions, and can be seen and heard as specified in Tables 23 and 24;

b. that systems react correctly following return to normal status;

c. that all indicator panels and switches are correctly marked;

d. that all functions on all indicator panels operate correctly;

e. that the system will operate from the essential supply stand-by power source;

f. that all indicator panels are labelled to show the areas they serve, or as detailed in the contract specifications.

15.90 The following tests should also be carried out:

a. for central indicator panels, check that the operation of the mute switch cancels the audible alarm and converts the flashing signals to steady, for all systems and conditions;

b. for repeater indicator panels, check that the mute switch cancels the audible alarm and that the flashing signals are converted to steady only on the central alarm panel, for all systems and conditions;

c. for area indicator panels, check that the operation of the mute switch cancels the audible only, for all systems and conditions;

d check power failure operates red "system fault" indicator and the audible alarm;

e. check that a contact line fault operates the "system fault" indicator, the main alarm displays and the audible alarm;

f. check audible reinstatement for each alarm panel;

g. check that the audible signal can be continuously muted via operation of the internal push-button for gas service alarm conditions only;

h. check for correct identification of each gas service on alarm panels and "departmental" or plant specifying labels;

j. check that each alarm panel emits the correct (two-tone) audible alarm. (Some manufacturers supply panels set for a single tone – in use, staff may confuse this sound with that emitted by some models of patient monitoring equipment.)

15.91 The results of the tests are recorded on Form B11.

## Verification of as-fitted drawings

15.92 The as-fitted drawings should be checked to ensure that all variations from the contract drawings have been recorded and the results may be recorded on Form B12.

## Filling with medical air

15.93 An indefinite time may elapse after completion of the MGPS before the system is taken into use. The installation contract may be written in the expectation that this will happen. In such circumstances the contract should require that the particulate contamination and odour tests specified in paragraphs 15.130–15.136 are carried out as an interim measure, using medical air as the test gas. Satisfactory completion of these particulate contamination and odour tests may then signify the completion of the construction contract.

15.94 It is the responsibility of the client to ensure that proper provision is made in a specific contract for the maintenance of the systems, their integrity, and any special connectors that may be required during this interim period.

15.95 All MGPS should be left filled with medical air at pipeline distribution pressure until they are filled with the specific working gas shortly before use. The medical vacuum pipeline need not be maintained under vacuum.

15.96 Provision should be made for regular running and maintenance of all supply plant during such an interim period.

15.97 Details of the work carried out, as well as records of the system pressures, should be recorded. This information is required in order to demonstrate that the systems have been satisfactorily maintained under pressure during this interim period. Tests for particulate contamination should be carried out after the systems are filled with the specific gas. The extent of the tests is at the discretion of the Quality Controller (MGPS).

15.98 The "Danger – do not use" label should remain affixed to each terminal unit until all testing is completed.

15.99 When the construction contract has finished, the contractor should record the removal of all special connectors and cylinders from site.

### Note

Special connectors and cylinders may be required to maintain the systems under pressure. This may be some time prior to the admission of patients. In such circumstances some contracts require systems to be completed and certificated for the purpose of practical completion and handover to the client.

## Purging and filling with specific gases

15.100 Each pipeline system must be purged with the specific working gas shortly before use. The following conditions should apply:

a. all sources of test gas must be disconnected;

b. all special connectors must be removed from site;

c. each pipeline system must be at atmospheric pressure with all AVSUs open;

d. each system must be filled to pipeline distribution pressure with the specific gas from the supply system;

e. with the supply system on, each terminal unit must be purged at a known flow with a volume of gas at least equal to the volume of the pipeline section being tested;

f. all oxygen, nitrous oxide, nitrous oxide/oxygen mixtures and helium/oxygen mixture discharged during the process must be released to a safe place.

15.101 The results of the purging process may be recorded on Form B13.

15.102 Purging is not necessary for vacuum systems.

### Note

It may be possible to carry out the tests outlined in paragraphs 15.57–15.90 with the working gases, either sequentially or consecutively, followed by the appropriate pharmaceutical test. After the tests, "certification" arrangements should be put in place such that the client takes over responsibility for managing the system.

## Pharmaceutical testing

15.103 When modifying existing systems, the test programme is agreed by the Quality Controller (MGPS) and Authorised Person (MGPS) and the system is taken back into use only after testing has been completed satisfactorily under a permit-to-work system.

15.104 Three types of work are identified:

a. new;

b. extension/upgrade; and

c. repair.

For new installations, for example a new ward, a new department, or a complete hospital, the Quality Controller (MGPS) will prepare a report containing details of tests carried out and an inventory of outlets tested.

15.105 For extensions, upgrades and repairs, the permit will provide the minimum report, although a longer report may be provided at the discretion of the Quality Controller (MGPS).

15.106 Inclusion of details such as the mounting order of terminal units observed at the time of test should be confirmed by signature of the Authorised Person (MGPS) on the report.

15.107 This inventory should be checked by the Authorised Person (MGPS) to ensure that all terminal units have been identified and tested. NIST connectors will also need to be identified. The Authorised Person (MGPS) should then amend a copy of the Quality Controller (MGPS) report, confirming in writing that the system can be taken into use.

15.108 Although a structured approach to testing should be adopted, access and time limitations to parts of the MGPS may lead to some disruption of proposed test regimes.

### Note

**Important:** These tests are described in the context of commissioning and repair/alteration to existing systems. However, it must be remembered that quarterly testing is also required by this Health Technical Memorandum and in accordance with Tables 28 and 29 for medical air generated by on-site air compressors, synthetic air from gas blending plant, and oxygen from oxygen concentrators using compressor plant. The Authorised Person (MGPS) should liaise with the Quality Controller (MGPS) to ensure that these tests are carried out and recorded.

## Notes

The role, responsibilities and relationships of Authorised Persons (MGPS) – in the context of both existing and new installations, where the Authorised Person (MGPS) may not have responsibility for the systems when in use – is covered in Part B.

Although precluded by adherence to the procedures recommended in this Health Technical Memorandum, there have been instances when a system undergoes further modifications after a Quality Controller (MGPS) has started testing. It is very important that any modifications are documented and that any additional testing required as a consequence of those modifications is performed. Use of the permit-to-work form should be considered in these cases.

If it is necessary to modify systems during or after completion of testing, the Authorised Person (MGPS) and Quality Controller (MGPS) will identify the extent of retesting that is required.

If the system has been completed and all documentation handed over to the client or operator of the building, any further modification must be carried out strictly in accordance with the permit-to-work procedure.

The importance of NIST connectors in facilitating engineering and pharmaceutical testing and their value during shutdowns and emergency situations should not be underestimated. All LVAs are fitted with upstream and downstream NIST connectors.

# Quality of medical gas systems

## General

15.109 The objective of these tests is to establish whether the pipeline has been contaminated during construction or modification. The tests indicate whether work has affected a gas, but they are not tests that indicate compliance with pharmaceutical specifications for licensed products.

15.110 Particulate contamination and odour tests may have been carried out prior to filling with the working gas, particularly if the system has been left for some time before use. However, the Quality Controller (MGPS) will define the extent of repetition of these tests, after the systems have been filled with the specific working gas.

15.111 The Quality Controller (MGPS) will also define the extent of all other pharmaceutical testing, depending on such factors as the extent and nature of the work and the age and condition of the existing systems.

15.112 The Ph. Eur. should be seen as a basic minimum standard when examining medical gas quality, as its principal application is to the manufacture and distribution of medicines according to well-established manufacturing processes. It is not intended to deal with the endless possibilities for contamination that are introduced by an MGPS, or the types of failure that might occur with on-site generation of gases.

15.113 Occasionally, the Quality Controller (MGPS) may need to resort to more sophisticated testing than is permitted by the use of portable equipment.

15.114 Oxygen, helium/oxygen mixture, nitrous oxide, and nitrous oxide/oxygen mixtures discharged during the process must be released to a safe place. These tests are not required on a vacuum system for any work including modifications or new works.

15.115 These test procedures are based on existing practice. The particulate contamination test is subjective in that it requires the QC to make a judgement on whether or not particles are visible on the filter.

15.116 The oil, water, carbon monoxide and carbon dioxide, sulphur dioxide and oxides of nitrogen tests can be carried out with detector tubes, but advances in detection technology have produced a range of suitable alternative instruments. The use of detector tubes giving a quantitative response is recommended, but if other equipment is used for validation purposes, it must provide a level of repeatability, resolution and accuracy at least equivalent to that of detector tubes and must be calibrated to appropriate Standards.

15.117 An electronic dew-point meter should be used in preference to water content measurements.

15.118 Detector tubes give a quantitative response and are not intended for re-use. They should be agent-specific, since non-agent-specific (polytest) tubes can respond to various agents such as volatile inorganic compounds, giving misleading results.

15.119 However, it is recommended that the use of polytest tubes be considered as an optional general test for contamination of pipelines on a representative sample of terminal units.

15.120 Users must be aware of the limitations of all types of detection equipment, including ambient operating conditions and cross-sensitivities specified for each type of detector tube.

15.121 These tests must be carried out on a representative sample of terminal units/NISTs in each system at the discretion of the Quality Controller (MGPS).

15.122 If terminal units are being tested, the sample must include, as a minimum, the most distant terminal unit on each branch. This may be the first terminal unit to be tested, but care must be taken to ensure, for example, that a new extension connected to old pipework is first well purged via a terminal unit/NIST connector as close as possible to the junction of the systems so as to avoid the spread of any contamination into the extension.

15.123 It will not normally be necessary to test the most distant terminal unit if distal NIST connectors are provided.

15.124 Depending on the results of this test, the Quality Controller (MGPS) should decide the number and location of additional terminal units/NISTs to be tested.

## Note

Provision of NISTs throughout an MGPS is to be encouraged, as this will greatly facilitate testing, particularly QC testing.

15.125 These tests are summarised in Table 28.

15.126 All sources of supply should be tested for quality before the pipeline distribution system is filled with the working gas. This test is not intended as a test of certificated gases but is to ensure that supply source equipment (manifolds, compressors, VIEs etc) does not compromise the quality of such gases when delivering them to the pipeline systems.

15.127 When extending existing systems, supply sources will not normally be retested before being used to fill the extension with the working gases.

15.128 For new installations, quality tests should be carried out on the plant as well as on the pipeline distribution system.

15.129 The results of the test may be recorded on Form B14.

### Particulate matter

15.130 MGPS should be free from particulate contamination, as they have been constructed using chemically cleaned, capped components and joined in a controlled process using a filtered shield gas.

15.131 However, on-site contamination can occur from ingress of building materials, dust etc. The presence of such particles can adversely affect the quality of the delivered gases. Therefore, tests to indicate their absence are important.

15.132 New systems should be purged until the particulate filter is completely clear of visible particles when viewed in a good light.

15.133 Older systems may exhibit particulates, even after considerable purging, as they can be released or carried along by the gas stream after disruption of the system, reverse gas flow, pressure waves down the pipe, or physical vibration.

## Note

When connecting new pipework to an older (possibly contaminated) system it may be advisable to perform the first purge via the inlet NIST of the first AVSU, or the first terminal unit of the new system, in order to reduce the possible spread of contamination into the new system.

15.134 Where it is evident that extended purging may not completely clear the system of particulates, a decision to accept the level of contamination present, agree a cleansing procedure or, in very exceptional circumstances, condemn the system will be made.

15.135 The test for particulate matter should be carried out at every terminal unit on a new system. It can be carried out either after completion of the construction phase using medical air (see paragraph 15.93) or after the system has been filled with the specified gas. Once the system is filled with working gas, it would not normally be necessary to repeat the test at every terminal unit. The actual number of terminal units sampled is at the discretion of the Quality Controller (MGPS). It would, however, be necessary to repeat the test in full where there is insufficient evidence to show that the system has been satisfactorily maintained under operating pressure with medical air for the interim period.

15.136 When tested with a membrane filter at a flow not less than 150 L/min for 30 s, the filter must be free from visible particles when viewed in good light. A suitable test device is described in Appendix D.

## Notes

When testing surgical air terminal units, a flow of 350 L/min for 30 s should be used.

When testing nitrous oxide/oxygen mixture terminal units, a flow of 275 L/min for 30 s should be used.

When testing helium/oxygen mixture systems, oxygen-free nitrogen is used at the manifold and this will require special connectors.

When tests and/or purging is carried out with the sources of supply serving an operational hospital, it is essential to ensure that the test flows used are not detrimental to the continuity or adequacy of supply in operational areas. When a flow rate of 150 L/min or more may not be possible without compromising the hospital system, a lower flow rate should be used at the discretion of the Quality Controller (MGPS).

### Oil

15.137 This test should be carried out at the plant test point of all newly installed medical/surgical compressed air plant and for all medical/surgical compressed air plant on a quarterly basis.

15.138 When break-ins to a tested (and compliant) medical/surgical air system have been completed, repetition of this test will not normally be required.

15.139 Work involving strip-down of compressor plant and subsequent replacement of oil-sealing components may require a follow-up oil test, at the discretion of the Quality Controller (MGPS).

15.140 Oil may be present as liquid, aerosol or vapour, and an appropriate test device is described in Appendix E.

15.141 The total oil content should be in accordance with Table 28.

15.142 It is desirable to carry out this test at a plant test point before any pipeline system is supplied by that plant so as to prevent inadvertent contamination of the distribution system.

15.143 A representative sample of terminal units on both new and modified medical compressed air and oxygen concentrator systems supplied by compressor plant may be checked at the discretion of the Quality Controller (MGPS).

15.144 Care should be taken in siting the test point to ensure a representative sample.

### Water

15.145 This test is intended to identify contamination of the pipeline system by moisture. It should not be confused with the test for compressor plant dryer performance, although it may indicate a failure in the dryer system.

## Notes

When testing terminal units supplied via low pressure, flexible connecting assemblies, it is often found that – on initial testing – moisture levels exceed the 0.05 mg/L limit; this is the result of desorption of minute quantities of moisture into the gas stream. This is particularly noticeable where the test flow is low, and should not cause undue concern. The Quality Controller (MGPS) should establish, however, that the elevated readings at such terminal units result from this effect and not water contamination of the pipeline. (For example, the results should be compared with the readings achieved at nearby terminal units supplied by copper pipework.) New developments in hose materials may lead to hoses with reduced water vapour permeability characteristics.

The effects of flow rate through dryer units and sampling times on detection equipment indications should also be taken into account when measuring water content.

15.146 The plant test point and a representative sample of terminal units distributed throughout the pipeline systems should be tested for total water content. The water content must not exceed 67 vpm (equivalent to an atmospheric pressure dew-point of approximately –46°C). The typical water content of medical gas cylinders is normally below 5 vpm. Water vapour content may be measured using the appropriate test device described in Appendix E (see also paragraph 15.117).

## Note

Older compressor/dryer combinations may fail to meet the Ph. Eur. requirement of 67 vpm. In these circumstances, the Quality Controller (MGPS) will decide whether application of the older atmospheric dew-point limit of –40°C (127 vpm) is acceptable (Health Technical Memorandum 2022: 1997).

### Carbon monoxide

15.147 The most distant terminal units on each branch of a medical/surgical air pipeline system supplied from a compressor plant and PSA systems must be tested for carbon monoxide, although it would not normally be necessary to test more than five terminal units. The concentration of carbon monoxide should not exceed 5 ppm v/v. This may be measured at up to five terminal units in each system using the appropriate test devices described in Appendix E.

15.148 When break-ins to a tested (and compliant) medical/surgical air system have been completed, repetition of this test will not normally be required.

### Carbon dioxide

15.149 The most distant terminal unit on each branch of a medical/surgical air pipeline system supplied from a compressor or an oxygen concentrator plant must be tested for carbon dioxide.

15.150 The concentration of carbon dioxide must not exceed 500 ppm v/v in medical air or 300 ppm v/v in oxygen from an oxygen concentrator plant.

15.151 When break-ins to a tested (and compliant) medical/surgical air system have been completed, repetition of this test will not normally be required.

## Notes

Increasing or fluctuating carbon dioxide readings in air or PSA-generated oxygen can be an early indication of dryer failure or poor compressor maintenance.

Carbon dioxide is no longer used as an inert shield gas during pipeline brazing.

If carbon dioxide has been installed (see Chapter 11), the test methodology should be at the discretion of the Quality Controller (MGPS).

### Sulphur dioxide

15.152 The most distant terminal units in medical/surgical air pipeline systems supplied from a compressor plant, and oxygen terminal units supplied from a PSA plant, must be tested for sulphur dioxide. (It will not normally be necessary to test more than five terminal units in a single system.) The concentration should not exceed 1 ppm v/v.

15.153 When break-ins to a tested (and compliant) medical/surgical air system have been completed, repetition of this test (and those in paragraphs 15.111–15.114) will not normally be required.

### Oxides of nitrogen (NO and $NO_2$)

15.154 The most distant terminal units in medical/surgical air pipeline systems supplied from a compressor plant, and oxygen terminal units supplied from a PSA plant, must be tested for oxides of nitrogen. (It will not normally be necessary to test more than five terminal units in a single system.) The concentration should not exceed 2 ppm v/v.

15.155 When break-ins to a tested (and compliant) medical/surgical air system have been completed, repetition of this test (and those in paragraphs 15.111–15.114) will not normally be required.

**Important:**

See Note (d) in Table 29 on disparity between Ph. Eur. and EH40 with reference to acceptable levels of sulphur dioxide and oxides of nitrogen.

### Nitrogen

15.156 Oxygen-free nitrogen is used as the inert gas shield, and all terminal units of all gas systems should be tested to ensure that the systems have been adequately purged.

15.157 For oxygen systems and nitrous oxide/oxygen, an oxygen analyser must be used to ensure that the oxygen concentration is not less than that given in Table 30.

15.158 For nitrous oxide systems, an instrument based on thermal conductivity, or an infrared meter, must be used to check that the system has been adequately purged at every terminal unit.

15.159 If a thermal conductivity meter is used, it will be necessary to prove absence of carbon dioxide (which could have been used inadvertently as a shield gas) by the use of a chemical reagent tube.

### Pipeline odour/taste

15.160 An odour test is performed because it incorporates, qualitatively, many impurity checks, as several contaminants are detectable by odour. This test is normally carried out as the final test with the working gases, except for nitrous oxide, nitrous oxide/oxygen and carbon dioxide, which should not be inhaled. The odour threshold of particulate matter is approximately 0.3 $mg/m^3$.

15.161 In certain circumstances (see paragraph 15.93), it may be carried out as the first test after completion of construction of the pipeline installation using medical air as the test gas. In such circumstances, a pipeline odour/taste test can be carried out on nitrous oxide and nitrous oxide/oxygen systems.

15.162 In addition to all new terminal units, a representative sample of terminal units on existing parts of the systems must be checked.

## Notes

For some time it has been known that materials used in the construction of low-pressure connecting assemblies can present an odour. This was recognised in the 1984 version of BS 5682:

*"Plastics materials currently in use will release small quantities of volatile organic matter into the gas stream throughout the life of the plastics components of the material. The quantities released appear to be below the levels normally considered toxic but, as yet, insufficient research has been carried out to be able to identify and quantify these components, therefore no tests are recommended."*

More recent work has shown that the quantities of those agents that can be identified are significantly below levels considered to be toxic. It is possible that developments in hose material structure will result in the reduction of odour (and moisture retention).

The Quality Controller (MGPS) should perform additional oil and polytest analyses if indistinct odours are detected where flexible hoses are not involved.

New developments in hose materials may lead to hoses that are odour-free and do not leach chemicals into the gas stream.

## Table 29 Quality specifications for medical gas pipeline tests (working gases)

| Gas and source | Particulates | Oil | Water | CO | $CO_2$ | NO and $NO_2$ | $SO_2$ | Poly-test tube (Optional) | Odour |
|---|---|---|---|---|---|---|---|---|---|
| Oxygen from PSA plant | Free from visible particles in a 75 L sample | ≤0.1 mg/m$^3$ | ≤67 vpm (≤0.05 mg/L, atmospheric dew-point of –46°C) | ≤5 mg/m$^3$; ≤5 ppm v/v | ≤300 ppm v/v | ≤2 ppm v/v | ≤1 ppm v/v | No discoloration | None |
| Nitrous oxide | Free from visible particles in a 75 L sample | – | ≤67 vpm (≤0.05 mg/L, atmospheric dew-point of –46°C) | – | – | – | – | No discoloration | SAFETY Not performed |
| Nitrous oxide/ oxygen mixture | Free from visible particles in a 75 L sample | – | ≤67 vpm (≤0.05 mg/L, atmospheric dew-point of –46°C) | – | – | – | – | No discoloration | SAFETY Not performed |
| Medical and surgical air | Free from visible particles in a 75 L sample (for medical air) and 175 L sample (for surgical air) | ≤0.1 mg/m$^3$ | ≤67 vpm (≤0.05 mg/L, atmospheric dew-point of –46°C) | ≤5 mg/m$^3$; ≤5 ppm v/v | ≤900 mg/m$^3$; ≤500 ppm v/v | ≤2 ppm v/v | ≤1 ppm v/v | No discoloration | None |
| Dental compressed air | Free from visible particles in a 75 L sample | ≤0.1 mg/m$^3$ | ≤1020 vpm (≤0.78 mg/L, atmospheric dew-point of –20°C) | ≤5 mg/m$^3$; ≤5 ppm v/v | ≤900 mg/m$^3$; ≤500 ppm v/v | ≤2 ppm v/v | ≤1 ppm v/v | No discoloration | None |
| Synthetic air | Free from visible particles in a 75 L sample | – | ≤67 vpm (≤0.05 mg/L, atmospheric dew-point of –46°C) | – | – | – | – | No discoloration | None |
| Oxygen from bulk liquid or cylinders | Free from visible particles in a 75 L sample | – | ≤67 vpm (≤0.05 mg/L, atmospheric dew-point of –46°C) | – | – | – | – | No discoloration | None |
| Helium/ oxygen mixture $O_2$, <30% | Free from visible particles in a 75 L sample | – | ≤67 vpm (≤0.05 mg/L, atmospheric dew-point of –46°C) | – | – | – | – | No discoloration | None |

**Notes:**

a. The quality of the gases delivered at the terminal units should also comply with the specifications given in the current edition of the Ph. Eur. (see Table 30). Additionally, contamination introduced by the MGPS, and not limited by the Ph. Eur. specification, should not exceed levels that might pose a threat to patients. It should be borne in mind that the safe levels for medical gases delivered to patients are likely to be significantly lower than those permitted for healthy individuals.
In addition to the monograph, the official standards section of the general notices should be read.

b. The tests for oil, carbon monoxide, carbon dioxide, sulphur dioxide and oxides of nitrogen are not normally carried out when the source of supply is from cylinders or cryogenic systems, although it should be noted that rare instances of oil contamination arising from the pipeline have occurred.

c. Synthetic air will be tested for identity as shown in Table 30. A GLC (gas-liquid chromatography) test for nitrogen is possible but not without practical difficulties. Nitrogen content will, therefore, usually be inferred from oxygen analyser test results.

d. The Health and Safety Executive has revised its guidance on exposure limits for sulphur dioxide, nitrogen monoxide and nitrogen dioxide.
   Occupational exposure standards (OESs) for nitric oxide, nitrogen dioxide and sulphur dioxide were removed from EH40 in the Amendment of April 2003.
   Time-weighted averages (TWAs) for nitrogen monoxide and nitrogen dioxide are now suggested as no greater than 1 ppm, and limits for sulphur dioxide exposure as less than 1 ppm for both 8-hour TWA OES and 15-minute STEL (short-term exposure limit).
   Some breathing air standards seek to limit the levels of such contaminants to 10% of the 8-hour TWA, as medical gases are intended for use by people who are not in the best of health. Therefore it is suggested, when testing for these specific compounds, or any contaminants not listed in the Ph. Eur., that a limit of 10% of the OES should be confirmed.
e. See Note following paragraph 15.146.

## Table 30 Ph. Eur. quality specifications for medical gases

| Gas and source | Oil | Water | CO | $CO_2$ | NO and $NO_2$ | $SO_2$ | Odour/Taste |
|---|---|---|---|---|---|---|---|
| Oxygen from bulk liquid or cylinders | – | ≤67 vpm (≤0.05 mg/L, atmospheric dew-point of –46°C) | ≤5 mg/m$^3$; ≤5 ppm v/v | ≤300 ppm v/v | – | – | None |
| Oxygen from PSA plant | 0.1 mg/m$^3$ | ≤67 vpm (≤0.05 mg/L, atmospheric dew-point of –46°C) | ≤5 mg/m$^3$; ≤5 ppm v/v | ≤300 ppm v/v | ≤2 ppm v/v | ≤1 ppm v/v | None |
| Nitrous oxide | – | ≤67 vpm (≤0.05 mg/L, atmospheric dew-point of –46°C) | ≤5 mg/m$^3$; ≤5 ppm v/v | ≤300 ppm v/v | ≤2 ppm v/v | – | N/A |
| Nitrous oxide/oxygen mixture | – | ≤67 vpm (≤0.05 mg/L, atmospheric dew-point of –46°C) | ≤5 mg/m$^3$; ≤5 ppm v/v | ≤300 ppm v/v | ≤2 ppm v/v | – | N/A |
| Medical and surgical air | 0.1 mg/m$^3$ | ≤67 vpm (≤0.05 mg/L, atmospheric dew-point of –46°C) | ≤5 mg/m$^3$; ≤5 ppm v/v | ≤900 mg/m$^3$; ≤500 ppm v/v | ≤2 ppm v/v | ≤1 ppm v/v | None |
| Synthetic air | – | ≤67 vpm (≤0.05 mg/L, atmospheric dew-point of –46°C) | – | – | – | – | None |
| Helium/ oxygen mixture $O_2$, <30% | – | ≤67 vpm (≤0.05 mg/L, atmospheric dew-point of –46°C) | ≤5 mg/m$^3$; ≤5 ppm v/v | ≤300 ppm v/v | ≤2 ppm v/v | – | None |

### Note

Particulate level tests and polytests are NOT included in the Ph. Eur.

## Gas identification

15.163 The identity of the gas must be tested at terminal units on medical gas pipeline systems. This would include all new terminal units, whether on a new installation or a modification or extension, and a representative sample of terminal units on an existing system, which may have been affected by the work. All systems must have been filled with the specific gas according to paragraph 15.100.

15.164 The composition of all compressed gases must be positively identified. This can be accomplished using an oxygen analyser for oxygen, nitrous oxide/oxygen and air, and a thermal conductivity or infrared meter for nitrous oxide.

15.165 When checking the identity of nitrous oxide and nitrous oxide/oxygen mixture, the gas should be discharged in a manner that minimises pollution and personnel exposure.

15.166 When testing pipelines for helium/oxygen mixture, an initial test is carried out with nitrogen connected after completing the particulate test. An oxygen analyser is used and all terminal units are tested. After a zero reading is achieved, product cylinders are connected and the system is purged. A second test is performed with an oxygen analyser; the oxygen content should be as in Table 31.

15.167 The nominal gas concentration at specific terminal units is given in Table 31; vacuum must be identified by observation of suction at the terminal unit.

## Test results

15.168 The test results for quality and gas identity may be recorded on Form B15.

### Table 31 Gas concentrations for identification purposes

| Gas and source | Paramagnetic oxygen analyser reading | Thermal conductivity (TC)/infrared (IR) instrument reading | Carbon dioxide detector tube indication if TC meter used | Vacuum probe |
|---|---|---|---|---|
| Oxygen from liquid or cylinders | Minimum 99.5% | – | – | – |
| Oxygen from concentrator | Minimum 94.0% | – | – | – |
| Nitrous oxide | –0.2% | Indicates “nitrous oxide” or gives a reading of 100% ± 2.0% (TC), ≥98% (IR) | ≤300 ppm v/v | – |
| Nitrous oxide/oxygen mixture | 50.0% ± 2.0% | 50.0% ± 2.0% | – | – |
| Medical, surgical and dental air | 20.9% ± 0.5% | – | – | – |
| Synthetic air | 95–105% of nominal value of 21.0–22.5% | – | – | – |
| Vacuum | – | – | – | Suction present |
| Nitrogen shield gas | 0% | 0% (IR) | – | – |
| Helium/oxygen mixture:<br>Test 1.<br>Test 2. | <br><br>0%<br>20.9% ± 0.5% | <br><br>0%<br>20.9% ± 0.5% | – | – |

**Note:**

The tolerance of the measuring instrument should be allowed in addition. For oxygen concentrator plant (PSA) supplied system, the minimum concentration must be 94% oxygen. A vacuum gauge may be used to obtain a quantitative reading of vacuum level and verify terminal unit performance.

## AGS disposal systems

### General

15.169 BS 6834:1987 specifies the tests to be carried out on AGS disposal systems that comply with the British Standard. The tests specified are designed to ensure adequate performance and that the safety provisions of receiving systems will be met.

15.170 The responsibility for the tests should be clearly identified at the contract stage for new installations, in the same way as for the medical gas pipeline system. In general, the contractor should carry out the tests, which should be witnessed by the Authorised Person (MGPS).

### Performance tests

15.171 All equipment should be tested to ensure that it performs satisfactorily during continuous operation under full load for one hour.

15.172 All electrically powered equipment should be tested as follows:

- check for correct rotation;
- check the current through the powered device at full load.

15.173 The disposal system should be tested to ensure that it meets the requirements set out in the table below, with the number of terminal units for which it has been designed in use.

| | Disposal system standard | | | |
|---|---|---|---|---|
| | Pressure drop | | Flow rate | |
| | BS 6834: 1987 | ISO DIS 7396-2: 2005 | BS 6834: 1987 | ISO DIS 7396-2: 2005 |
| Maximum | 1 kPa | 1 kPa | 130 L/min | 80 L/min |
| Minimum | 4 kPa | 2 kPa | 80 L/min | 50 L/min |
| Maximum static pressure | 20 kPa (–ve) | 15 kPa (–ve) | This check is made before performing the flow tests | |

**Note**

Since the preparation of BS 6834:1987, developments in anaesthesiology have resulted in reduced flows being used. Depending on local circumstances, it may be possible to commission systems for different flows in accordance with ISO DIS 7396-2:2005. Details of the test flows should be recorded in the commissioning documentation.

15.174 The test should be carried out as described in Appendix K of BS 6834:1987. The test device is inserted into each terminal unit in turn and checked for pressure at flows of 80 L/min and 130 L/min for BS systems, and 50 L/min and 80 L/min for ISO systems. Adjustment is then made if necessary.

15.175 The test device and a number of metered leaks are then inserted into the system to replicate the design flow. The measurements above are repeated. If the test results are satisfactory, the test device is removed and substituted by a metered leak.

**Note**

The test device is designed to replicate either type of receiving system for which the disposal system has been designed (see paragraph 15.73).

15.176 The other terminal units are then tested in turn by substituting the test device for each metered leak including the test device.

**Note**

For the purposes of diversity, it may be assumed that in any operating department only one receiving system for each operating room is in use at any time. (In a typical theatre suite with two terminal units in the operating room and one in the anaesthetic room, the total number of metered leaks used for testing is two; that is, one being placed in an operating room terminal unit and the other in the anaesthetic room terminal unit.)

15.177 The operation of user-controlled switches, power-on indicators and alarm systems should also be checked.

15.178 AGSS terminal units should be checked for correct mechanical operation and that the check valve operates satisfactorily.

## Requirements before a medical gas pipeline system is taken into use

### General

15.179 Before a system is used, the appropriate persons must certify in writing that the tests and procedures required in paragraphs 15.46–15.100 and 15.109–15.178 have been completed, and that all systems comply with the requirements.

This must include certification that all drawings and manuals required by the contract have been supplied and as-fitted drawings are correct (see Form B16).

**15.180** All certificates must be dated and signed by the appropriate witnesses, by the contract-supervising officer and by the representative of the contractor.

**15.181** For modifications or extensions to existing systems, the performance tests for flow and pressure drop (as described in paragraphs 15.41–15.43) should be carried out on the completed system if practicable. If the performance is in accordance with the specification prepared (as described in paragraphs 15.32–15.44), the system may be taken into use, provided that all the other tests have been satisfactorily completed.

### Note

In many cases, the extension/modification will be relatively small and unlikely to significantly affect the performance of the system.

## Operational policy

**15.182** A procedure must be available in accordance with Part B, and must ensure continuity of supply of cylinders and bulk liquid. This will incorporate a procedure for recording delivery, handling and storage of full and empty cylinders, with an indication of who is responsible for these activities. The supplier must certify the composition of the cylinder contents. All deliveries of bulk liquid oxygen should be tested for conformance to the product licence specification before dispatch by the supplier, and should be supplied with a certificate indicating compliance.

## Cylinder storage and handling

**15.183** There should be recorded visual checks for correct labelling, including batch numbers (see Part B).

## Removal of construction labels

**15.184** When all tests have been completed satisfactorily, the "Danger – do not use" labels affixed to terminal units should be removed on the authority of, or by, the Authorised Person (MGPS).

# Appendix A – Testing, commissioning and filling for use: forms to be completed during testing and commissioning of piped medical gases systems

| | Form |
|---|---|
| Summary of tests | B0 |
| **Carcass tests** | |
| Labelling and marking | B1 |
| Sleeving and supports | B1 |
| Leakage test | B1 |
| Cross-connection test | B2 |
| **System tests** | |
| Leakage test | B3 |
| Vacuum leakage test | B4 |
| AVSUs – closure and zoning tests | B5A |
| LVAs – closure and zoning tests | B5B |
| Cross-connection test | B6 |
| Functional test of terminal units | B7A |
| Functional tests of NIST connectors | B7B |
| Design flow performance test | B8 |
| Functional test of supply system | B9 |
| Pressure safety valves | B10 |
| Warning systems | B11 |
| Verification of drawings | B12 |
| Purging and filling | B13 |
| Quality | B14 |
| Gas identification | B15 |
| Certificate of completion | B16 |

## Medical Gas Pipeline Carcass Tests

**Form B0 (Sheet of Sheets)**

Hospital ................................................ Scheme ................................................

File Number ................................................ Date ................................................

**Summary of Tests**

This is to certify that the following tests have been carried out:

| System | Form | Test carried out satisfactorily |
|---|---|---|
| **Carcass tests** | | |
| Labelling and marking | B1 | |
| Sleeving and supports | B1 | |
| Leakage test | B1 | |
| Cross-connection test | B2 | |
| **System tests** | | |
| Leakage test | B3 | |
| Vacuum leakage test | B4 | |
| Area valve service units – closure and zoning test | B5A | |
| Line valve assemblies | B5B | |
| Cross-connection test | B6 | |
| Functional tests of terminal units | B7A | |
| Functional tests of NIST connectors | B7B | |
| Design flow performance tests | B8 | |
| Sources of supply | B9 | |
| Pressure safety valves | B10 | |
| Warning systems | B11 | |
| Verification of drawings | B12 | |
| Purging and filling | B13 | |
| Quality | B14 | |
| Gas identification | B15 | |
| Permit-to-work form | | |
| Construction labels removed | | |

Contract Supervising Officer

Status ................................................ Signed ................................................

Date ................................................ Name ................................................

All appropriate tests satisfactorily carried out. System may now be taken into use.

Authorised Person (MGPS)

Status ................................................ Signed ................................................

Date ................................................ Name ................................................

## Medical Gas Pipeline Carcass Tests

Form B1 (Sheet of Sheets)

Hospital ........................................................................................ Scheme ............................................................................

File Number .................................................................................. Date .................................................................................

**Part 1 – Leakage test, labelling and marking, sleeving and supports**

This is to certify that a LEAKAGE test in accordance with paragraphs 15.49–15.51 was carried out on the piped system on this scheme and that during the test, a pressure, as shown in column 2 below, was held as follows. A certified gauge number was used.

| Section tested | Test pressure (kPa) | Hours on test | Pressure drop (kPa) | Pressure loss (kPa/h) | Pass/Fail <0.2/2h (400 kPa systems) <0.5/2h (700 kPa systems) |
|---|---|---|---|---|---|
| | | | | | |
| | | | | | |
| | | | | | |
| | | | | | |
| | | | | | |
| | | | | | |
| | | | | | |
| | | | | | |
| | | | | | |
| | | | | | |

**Part 2 – Links between systems**

For the purpose of carrying out this test, the following links have been made:

..........................................................................................................................................................................................

..........................................................................................................................................................................................

This is to certify that the above tests have been carried out and that the following links have been removed:

..........................................................................................................................................................................................

..........................................................................................................................................................................................

Contractor's Representative

Status ............................................................................................ Signed ............................................................................

Date ............................................................................................. Name .............................................................................

Contract Supervising Officer

Status ............................................................................................ Signed ............................................................................

Date ............................................................................................. Name .............................................................................

## Medical Gas Pipeline Carcass Tests

**Form B2 (Sheet of Sheets)**

Hospital .................................................................................. Scheme ..................................................................................

File Number .......................................................................... Date ......................................................................................

**Cross-connection test**

This is to certify that a CROSS-CONNECTION test in accordance with paragraphs 15.52–15.55 was carried out on the following medical gas pipeline systems:

...........................................................................................................................................................................................

...........................................................................................................................................................................................

...........................................................................................................................................................................................

...........................................................................................................................................................................................

...........................................................................................................................................................................................

...........................................................................................................................................................................................

...........................................................................................................................................................................................

...........................................................................................................................................................................................

...........................................................................................................................................................................................

...........................................................................................................................................................................................

...........................................................................................................................................................................................

...........................................................................................................................................................................................

...........................................................................................................................................................................................

...........................................................................................................................................................................................

...........................................................................................................................................................................................

...........................................................................................................................................................................................

...........................................................................................................................................................................................

...........................................................................................................................................................................................

...........................................................................................................................................................................................

...........................................................................................................................................................................................

...........................................................................................................................................................................................

...........................................................................................................................................................................................

...........................................................................................................................................................................................

...........................................................................................................................................................................................

...........................................................................................................................................................................................

...........................................................................................................................................................................................

...........................................................................................................................................................................................

No cross-connections between these systems were found.

Contractor's Representative

Status .................................................................................. Signed ..................................................................................

Date .................................................................................... Name ...................................................................................

## Medical Gas Pipeline Total System Tests

**Form B3 (Sheet of Sheets)**

Hospital .................................................... Scheme ....................................................

File Number .................................................... Date ....................................................

**Leakage test from total compressed gas system**

This is to certify that a LEAKAGE test in accordance with paragraphs 15.59–15.60 was carried out on the piped system on this scheme and that during the test, a pressure of was held for hours with a pressure drop of kPa.

| Section tested | Test pressure (kPa) | Hours on test | Pressure drop (kPa) | Pressure loss (kPa/h) | Pass/Fail <0.2/2h (400 kPa systems) <0.5/2h (700 kPa systems) |
|---|---|---|---|---|---|
| | | | | | |
| | | | | | |
| | | | | | |
| | | | | | |
| | | | | | |
| | | | | | |
| | | | | | |
| | | | | | |
| | | | | | |
| | | | | | |
| | | | | | |
| | | | | | |

Contractor's Representative

Status .................................................... Signed ....................................................

Date .................................................... Name ....................................................

Contract Supervising Officer

Status .................................................... Signed ....................................................

Date .................................................... Name ....................................................

Witnessed on behalf of ........................................................................................................

By .................................................... Status ....................................................

Signed .................................................... Date ....................................................

## Medical Gas Pipeline Total System Tests

**Form B4 (Sheet of Sheets)**

Hospital ........................................................................................ Scheme ............................................................................

File Number ................................................................................. Date ...................................................................................

**Leakage into total vacuum system test**

This is to certify that a LEAKAGE test in accordance with paragraph 15.61 was carried out on the piped vacuum system at a system pressure of kPa. The pressure increase after 1 hour was kPa (max 10 kPa).

Contractor's Representative

Status ......................................................................................... Signed ..............................................................................

Date ........................................................................................... Name ...............................................................................

Contract Supervising Officer

Status ......................................................................................... Signed ..............................................................................

Date ........................................................................................... Name ...............................................................................

Witnessed on behalf of ...........................................................................................................................................................

By ............................................................................................... Status ..............................................................................

Signed ......................................................................................... Date ................................................................................

## Medical Gas Pipeline Total System Tests

**Form B5A (Sheet of Sheets)**

Hospital ............................................................................ Scheme ............................................................................

File Number ............................................................................ Date ............................................................................

**Area Valve Service Units – closure and zoning tests**

This is to certify that CLOSURE and ZONING of the AVSUs was tested in accordance with paragraphs 15.63–15.68 on the pipeline system as follows:

| AVSU number | Test pressure (kPa) | Downstream/upstream pressure change after 15 min (kPa) | Terminal units controlled (total no) | Terminal unit labelling |
|---|---|---|---|---|
| | | | | |
| | | | | |
| | | | | |
| | | | | |
| | | | | |
| | | | | |
| | | | | |
| | | | | |
| | | | | |
| | | | | |
| | | | | |
| | | | | |
| | | | | |

Contractor's Representative

Status ............................................................................ Signed ............................................................................

Date ............................................................................ Name ............................................................................

Contract Supervising Officer

Status ............................................................................ Signed ............................................................................

Date ............................................................................ Name ............................................................................

Witnessed on behalf of ............................................................................

By ............................................................................ Status ............................................................................

Signed ............................................................................ Date ............................................................................

## Medical Gas Pipeline Total System Tests

**Form B5B (Sheet of Sheets)**

Hospital ........................................ Scheme ........................................

File Number ........................................ Date ........................................

**Line Valve Assemblies – closure and zoning tests**

This is to certify that CLOSURE and ZONING of the LVAs was tested in accordance with paragraphs 15.63–15.65 on the pipeline system as follows:

| LVA number | Test pressure (kPa) | Downstream/upstream pressure change after 15 min (kPa) | Terminal units controlled (total no) | Terminal unit labelling |
|---|---|---|---|---|
| | | | | |
| | | | | |
| | | | | |
| | | | | |
| | | | | |
| | | | | |
| | | | | |
| | | | | |
| | | | | |
| | | | | |
| | | | | |
| | | | | |
| | | | | |

Contractor's Representative

Status ........................................ Signed ........................................

Date ........................................ Name ........................................

Contract Supervising Officer

Status ........................................ Signed ........................................

Date ........................................ Name ........................................

Witnessed on behalf of ........................................

By ........................................ Status ........................................

Signed ........................................ Date ........................................

## Medical Gas Pipeline Total System Tests

**Form B6 (Sheet of Sheets)**

Hospital ............................................................ Scheme ............................................................

File Number ............................................................ Date ............................................................

**Cross-connection test**

This is to certify that a CROSS-CONNECTION test in accordance with paragraphs 15.70–15.74 was carried out on the following medical gas pipeline systems:

..........................................................................................................................................

..........................................................................................................................................

..........................................................................................................................................

..........................................................................................................................................

..........................................................................................................................................

..........................................................................................................................................

..........................................................................................................................................

..........................................................................................................................................

..........................................................................................................................................

..........................................................................................................................................

..........................................................................................................................................

..........................................................................................................................................

..........................................................................................................................................

..........................................................................................................................................

..........................................................................................................................................

..........................................................................................................................................

..........................................................................................................................................

..........................................................................................................................................

..........................................................................................................................................

..........................................................................................................................................

..........................................................................................................................................

..........................................................................................................................................

..........................................................................................................................................

Contractor's Representative

Status ............................................................ Signed ............................................................

Date ............................................................ Name ............................................................

Contract Supervising Officer

Status ............................................................ Signed ............................................................

Date ............................................................ Name ............................................................

Witnessed on behalf of ..........................................................................................................

By ............................................................ Status ............................................................

Signed ............................................................ Date ............................................................

## Medical Gas Pipeline Total System Tests

**Form B7A (Sheet of Sheets)**

Hospital ............................................................................................ Scheme ............................................................................

File Number .................................................................................. Date .................................................................................

**Functional tests of terminal units**
(in accordance with the contract specification and paragraphs 15.77–15.78)

System ............................................................................................................................

Specified flow ................................ L/min Specified pressure drop ................................ kPa

| Terminal unit number | Room number | Specified flow achieved Yes/No | Specified pressure drop achieved Yes/No | Mechanical function | Gas specificity |
|---|---|---|---|---|---|
| | | | | | |
| | | | | | |
| | | | | | |
| | | | | | |
| | | | | | |
| | | | | | |
| | | | | | |
| | | | | | |
| | | | | | |
| | | | | | |
| | | | | | |
| | | | | | |
| | | | | | |

Contractor's Representative

Status ............................................................................................ Signed ............................................................................

Date ............................................................................................ Name ............................................................................

Contract Supervising Officer

Status ............................................................................................ Signed ............................................................................

Date ............................................................................................ Name ............................................................................

Witnessed on behalf of ..............................................................................................................................................

By ............................................................................................ Status ............................................................................

Signed ............................................................................................ Date ............................................................................

**Medical Gas Pipeline Total System Tests** **Form B7B (Sheet of Sheets)**

Hospital ........................................................................................ Scheme ..............................................................................

File Number ................................................................................. Date ....................................................................................

**Functional tests NIST connectors**
(in accordance with the contract specification and paragraphs 15.80–15.81)

System ........................................................................................................................

| NIST gas | Location or identification | Room number | Gas specificity Pass/Fail | Self-sealing Adequate/Inadequate |
|---|---|---|---|---|
| | | | | |
| | | | | |
| | | | | |
| | | | | |
| | | | | |
| | | | | |
| | | | | |
| | | | | |
| | | | | |
| | | | | |
| | | | | |
| | | | | |
| | | | | |
| | | | | |

Contractor's Representative

Status ........................................................................................ Signed ..............................................................................

Date ........................................................................................... Name ...............................................................................

Contract Supervising Officer

Status ........................................................................................ Signed ..............................................................................

Date ........................................................................................... Name ...............................................................................

Witnessed on behalf of ..........................................................................................................................................

By ............................................................................................... Status ..............................................................................

Signed ........................................................................................ Date ................................................................................

## Medical Gas Pipeline Total System Tests

**Form B8 (Sheet of Sheets)**

Hospital ........................................ Scheme ........................................

File Number ........................................ Date ........................................

**Design flow performance**
(in accordance with paragraphs 15.83–15.84)

System gas ........................................ System design flow ................................ (L/min)

| Total number of terminal units in system at test flow(s) of: | | | | | | |
|---|---|---|---|---|---|---|
| 40 L/min | 80 L/min | 100 L/min | 275 L/min | 350 L/min | Total flow L/min | Single point test flows (Pass/Fail) |
| | | | | | | |

Contractor's Representative

Status ........................................ Signed ........................................

Date ........................................ Name ........................................

Contract Supervising Officer

Status ........................................ Signed ........................................

Date ........................................ Name ........................................

Witnessed on behalf of ........................................

By ........................................ Status ........................................

Signed ........................................ Date ........................................

## Medical Gas Pipeline Total System Tests

**Form B9 (Sheet of Sheets)**

Hospital ........................................ Scheme ........................................

File Number ........................................ Date ........................................

**Functional tests of supply systems**

This is to certify that the following sources of supply have been tested in accordance with paragraph 15.85 and the attached sheets and found to comply with the specification.

| Source of supply | Contractor's Representative Name/Signature | Contract Supervising Officer Name/Signature |
|---|---|---|
| Manifold | | |
| Manifold | | |
| Manifold | | |
| | | |
| | | |
| Liquid oxygen plant | | |
| | | |
| | | |
| | | |
| Air compressor | | |
| | | |
| | | |
| | | |
| Vacuum plant | | |
| | | |
| | | |
| | | |
| Oxygen concentrator | | |
| | | |
| | | |
| | | |

Witnessed on behalf of ........................................

By ........................................ Status ........................................

Signed ........................................ Date ........................................

## Medical Gas Pipeline Total System Tests

**Form B10 (Sheet of Sheets)**

Hospital ............................................................ Scheme ............................................................

File Number ............................................................ Date ............................................................

**Pressure safety valves**

The pressure safety valves fitted to the pipeline systems have been inspected together with their certification and are in accordance with the contract specification and paragraphs 15.86–15.88.

| Location | Valve number | Position | Pipeline distribution pressure (A) | Certified discharge pressure (B) | B/A (%) |
|---|---|---|---|---|---|
| | | | | | |
| | | | | | |
| | | | | | |
| | | | | | |
| | | | | | |
| | | | | | |
| | | | | | |
| | | | | | |
| | | | | | |
| | | | | | |
| | | | | | |
| | | | | | |
| | | | | | |

If certificates are not provided, do not sign.

Contractor's Representative

Status ............................................................ Signed ............................................................

Date ............................................................ Name ............................................................

Contract Supervising Officer

Status ............................................................ Signed ............................................................

Date ............................................................ Name ............................................................

Witnessed on behalf of ............................................................

By ............................................................ Status ............................................................

Signed ............................................................ Date ............................................................

## Medical Gas Pipeline Total System Tests

**Form B11 (Sheet of Sheets)**

Hospital ........................................ Scheme ........................................

File Number ........................................ Date ........................................

**Warning systems**

This is to certify that WARNING SYSTEMS on the following medical gas pipeline systems have been tested in accordance with paragraphs 15.89–15.91 as follows:

| System | $O_2$ | $N_2O$ | $N_2O/O_2$ | MA-4 | Surgical air | VAC |
|---|---|---|---|---|---|---|
| Specified warrning pressure | | | | | | |
| Observed warning pressure | | | | | | |
| Warning given | | | | | | |
| Return to normal | | | | | | |
| Marking | | | | | | |
| All functions on all stations | | | | | | |
| Stand-by power | | | | | | |
| | | | | | | |
| | | | | | | |
| | | | | | | |
| | | | | | | |
| | | | | | | |
| | | | | | | |
| | | | | | | |
| | | | | | | |
| | | | | | | |
| | | | | | | |
| | | | | | | |
| | | | | | | |
| | | | | | | |

Contractor's Representative

Status ........................................ Signed ........................................

Date ........................................ Name ........................................

Contract Supervising Officer

Status ........................................ Signed ........................................

Date ........................................ Name ........................................

Witnessed on behalf of ........................................

By ........................................ Status ........................................

Signed ........................................ Date ........................................

## Medical Gas Pipeline Total System Tests — Form B12 (Sheet of Sheets)

Hospital ............................................................ Scheme ............................................................

File Number ............................................................ Date ............................................................

**Verification of drawings**

This is to certify that in accordance with paragraph 15.92, the as-fitted drawings of the following systems record all variations from the contract drawings:

| System | Drawing numbers | Contractor's Representative Status/Name | Contract Supervising Officer Status/Name | Date |
|---|---|---|---|---|
| $O_2$ | | | | |
| $N_2O$ | | | | |
| $N_2O/O_2$ | | | | |
| MA-4 | | | | |
| Surgical air | | | | |
| VAC | | | | |
| AGS | | | | |
| Helium/oxygen mixture | | | | |
| Carbon dioxide | | | | |

Witnessed on behalf of ............................................................

By ............................................................ Status ............................................................

Signed ............................................................ Date ............................................................

## ·line Total System Tests Form B13 (Sheet of Sheets)

.................................................................... Scheme ............................................................................

.............................................. Date ..................................................................................

:ems have been purged and filled with medical air/$O_2$ free nitrogen/the working gases (delete
.ragraphs 15.93–15.99 and/or 15.100–15.101 as follows:

| | $O_2$ | $N_2O$ | $N_2O/O_2$ mixture | MA-4 | Surgical air | VAC | $H_2/O_2$ mixture | $CO_2$ |
|---|---|---|---|---|---|---|---|---|
| | | | | | | | | |
| | | | | | | | | |
| | | | | | | | | |
| | | | | | | | | |
| | | | | | | | | |
| | | | | | | | | |

Neil Fleming
David Guy
Michael Howa
Tom Jenner
Reggie Leek
Aiden McDonnell
Damien Mills
Sean Owens
Ian Phili

Co..

Status ......................... ............................................ Signed ..............................................................................

Date ................................................................................................ Name ..............................................................................

Contract Supervising Officer

Status ................................................................................................ Signed ..............................................................................

Date ................................................................................................ Name ..............................................................................

Witnessed on behalf of ..................................................................................................................................................

By ................................................................................................ Status ..............................................................................

Signed ................................................................................................ Date ..............................................................................

## Medical Gas Pipeline Total System Tests

**Form B14 (Sheet of Sheets)**

Hospital ............................................ Scheme ............................................

File Number ............................................ Date ............................................

Quality specifications for medical gas piepline tests (working gases). This is to certify that medical gas systems have been tested in accordance with paragraphs 15.109–15.162 as follows:

| Gas and source | Particulates | Oil | Water | CO | $CO_2$ | NO and $NO_2$ | $SO_2$ | Poly-test tube (Optional) | Odour | Tick when parameters are met |
|---|---|---|---|---|---|---|---|---|---|---|
| Oxygen from PSA plant | Free from visible particles in a 75 L sample | ≤0.1 mg/m³ | ≤67 vpm (≤0.05 mg/L, atmospheric dew-point of –46°C) | ≤5 mg/m³; ≤5 ppm v/v | ≤300 ppm v/v | ≤2 ppm v/v | ≤1 ppm v/v | No discoloration | None | |
| Nitrous oxide | Free from visible particles in a 75 L sample | – | ≤67 vpm (≤0.05 mg/L, atmospheric dew-point of –46°C) | – | – | – | – | No discoloration | SAFETY Not performed | |
| Nitrous oxide/ oxygen mixture | Free from visible particles in a 75 L sample | – | ≤67 vpm (≤0.05 mg/L, atmospheric dew-point of –46°C) | – | – | – | – | No discoloration | SAFETY Not performed | |
| Medical and surgical air | Free from visible particles in a 75 L sample (for medical air) and 175 L sample (for surgical air) | ≤0.1 mg/m³ | ≤67 vpm (≤0.05 mg/L, atmospheric dew-point of –46°C) | ≤5 mg/m³; ≤5 ppm v/v | ≤900 mg/m³; ≤500 ppm v/v | ≤2 ppm v/v | ≤1 ppm v/v | No discoloration | None | |
| Dental compressed air | Free from visible particles in a 75 L sample | ≤0.1 mg/m³ | ≤1020 vpm (≤0.78 mg/L, atmospheric dew-point of –20°C) | ≤5 mg/m³; ≤5 ppm v/v | ≤900 mg/m³; ≤500 ppm v/v | ≤2 ppm v/v | ≤1 ppm v/v | No discoloration | None | |
| Synthetic air | Free from visible particles in a 75 L sample | – | ≤67 vpm (≤0.05 mg/L, atmospheric dew-point of –46°C) | – | – | – | – | No discoloration | None | |
| Oxygen from bulk liquid or cylinders | Free from visible particles in a 75 L sample | – | ≤67 vpm (≤0.05 mg/L, atmospheric dew-point of –46°C) | – | – | – | – | No discoloration | None | |
| Helium/ oxygen mixture $O_2$, <30% | Free from visible particles in a 75 L sample | – | ≤67 vpm (≤0.05 mg/L, atmospheric dew-point of –46°C) | – | – | – | – | No discoloration | None | |

Contractor's Representative

Status ............................................ Signed ............................................

Date ............................................ Name ............................................

Contract Supervising Officer

Status ............................................ Signed ............................................

Date ............................................ Name ............................................

Quality Controller

Status ............................................ Signed ............................................

Date ............................................ Name ............................................

Witnessed on behalf of ............................................

By ............................................ Status ............................................

Signed ............................................ Date ............................................

## Medical Gas Pipeline Total System Tests — Form B15 (Sheet of Sheets)

Hospital ........................................ Scheme ........................................

File Number ........................................ Date ........................................

**Identification of medical gas pipeline working gases**

This is to certify that medical gas systems have been tested in accordance with paragraphs 15.163–15.167 and the results are as follows (insert values for gases – tick for vacuum):

| Gas and source | Paramagnetic oxygen analyser reading | Thermal conductivity/ infra-red instrument reading | Carbon dioxide detector tube indication if TC meter used | Vacuum probe |
|---|---|---|---|---|
| Oxygen from liquid or cylinders | | – | – | – |
| Oxygen from concentrator | | – | – | – |
| Nitrous oxide | | | | – |
| Nitrous oxide/oxygen mixture | | | – | – |
| Medical, surgical and dental air | | – | – | – |
| Synthetic air | | – | – | – |
| Vacuum | – | – | – | |
| Nitrogen shield gas | | | – | – |
| Helium/oxygen mixture<br>Test 1<br>Test 2 | | | – | – |

Contractor's Representative

Status ........................................ Signed ........................................

Date ........................................ Name ........................................

Contract Supervising Officer

Status ........................................ Signed ........................................

Date ........................................ Name ........................................

Quality Controller

Status ........................................ Signed ........................................

Date ........................................ Name ........................................

Witnessed on behalf of ........................................

By ........................................ Status ........................................

Signed ........................................ Date ........................................

## Certificate of Completion

**Form B16**

Hospital .................................................................................... NHS Trust ......................................................................

Medical Gas Installations – Location ..............................................................................................................................

This is to confirm that the following tests have been performed:

Mechanical functions tests

B1 to B11 inclusive

Quality and gas identity tests

B13 to B15 inclusive

in accordance with Health Technical Memorandum 02-01 Part A, Chapter 15, and that the results are satisfactory.

Signed ...........................................................................................

Quality Controller (MGPS)

Signed ...........................................................................................

Contractor's Representative (MGPS)

Signed ...........................................................................................

Authorised Person (MGPS) on behalf of Trust/other

Witnessed .......................................................................................

Trust/other Building operator/owner

Date ...............................................................................................

We, Trust/FM Contractor

accept responsibility for the systems above and undertake to carry out any future work in accordance with the recommendations of Health Technical Memorandum 02-01 and the permit-to-work procedures.

Signed ...........................................................................................

Date ...............................................................................................

# Appendix B – Gas pressure variation with temperature

## General

1. Tests are specified for leakage of the pipeline carcass and the pipeline systems. During these tests, pressure changes may occur that are caused by temperature changes rather than leakage.

2. Pressure changes due to temperature difference may be calculated according to the Gas Laws (see the 'Glossary' in Part B).

3. It is assumed that the temperature in the pipeline is uniform in all branches. If substantial runs are external, an average temperature should be chosen.

## Calculation

4. The change in gas pressure with temperature is as follows:

   $P1/T2 = P2/T1$

   where:

   $P1$ = the initial absolute pressure of a fixed volume of gas;

   $P2$ = the final absolute pressure of a fixed volume of gas;

   $T1$ = the initial absolute temperature;

   $T2$ = the final absolute temperature.

   Therefore:

   $P2 = (P1 \times T2)/T1.$ (1)

5. Care must be taken to express pressure and temperature in absolute values.

6. Pressure is normally expressed in "gauge" pressure:

   Absolute pressure = gauge pressure + atmospheric pressure.

7. Temperature is normally expressed in °C.

## Examples

8. The carcass of a medical air pipeline is tested for leakage at a working pressure of 14.0 bar pressure. The temperature is 13°C at the beginning of the test and 17°C at the end of the test:

   $P1 = 14.0 + 1.0 = 15.0$ bar

   $T1 = 273 + 13 = 286$ K

   $T2 = 273 + 17 = 290$ K.

9. Therefore, using Equation (1):

   $P2 = (15 \times 290)/286$
   $= 15.21$ bar (absolute pressure)
   $= 14.21$ bar (gauge pressure).

10. That is, gauge pressure should read 14.21 bar at the end of the test, assuming that no leakage has occurred.

# Appendix C – Pressure-drop test device

## General

1. Special test devices are required to measure the pressure at specified flows at each terminal unit.

2. Suitable test devices are commercially available or may be constructed in accordance with the outline specification given below.

## Measurement principle

3. Flow at a specified pressure may be measured either with a calibrated orifice or with a flowmeter.

4. Pressure may be measured with a bourdon gauge.

5. A gas-specific probe conforming to BS 5682:1998 should be used to connect the device to the terminal unit.

6. The test device is connected to the terminal unit by the gas-specific probe and the pressure at the specified flow is read on the gauge.

## Functional requirements

7. The test device should consist of the following components:

   - gas-specific probe to BS 5682:1998;
   - body on/off valve pressure gauge;
   - orifice or flowmeter.

8. The body may be of a design that allows exchange of the following components:

   a. gas-specific probes;

   b. calibrated orifices;

   c. pressure gauges.

9. An on/off valve may be incorporated into the body.

10. The complete assembly should be tested for leaks.

## Orifices

11. The orifices should be selected from the information on the manufacturer's data sheets or from practical testing.

12. These devices should be checked against a flowmeter before use.

## Flowmeter

13. A bobbin flowmeter calibrated to a flow of 40 L/min may be used to measure flow under vacuum.

## Pressure gauge

14. A 50 mm bourdon gauge with an appropriate full scale reading and interval should be used as follows:

| Test pressure kPa | Scale | Scale interval |
|---|---|---|
| 400 | 0–7 bar | 0.1 bar |
| 700 | 0–11 bar | 0.5 bar |
| Vacuum | 0–100 kPa (0–760 mm Hg) | 5 kPa (50 mm Hg) |
| 1 bar = 100 kPa approximately | | |

### Note

Generally it is found that three separate test devices for 400 kPa, 700 kPa and vacuum provide greater convenience. Because the test methodology in Chapter 15 has the potential for exposing the 400 kPa and vacuum devices to pressure or vacuum, it is desirable that they include an appropriate directional check valve.

# Appendix D – Membrane filter test device

## General

1. The function of this test device is to collect particulate material which may be present in the pipeline.

2. Filter holders appropriate to the pressure encountered are commercially available.

3. The filter holder should be specified for use at pipeline-distribution pressure and be oxygen-compatible.

## Measurement principle

4. A known volume of gas is passed through a membrane filter that will collect all visible particles.

5. Hydrophobic membrane filters of pore size 0.45 μm should be used.

## Test equipment

6. The following equipment is required:

   a. a membrane filter holder;

   b. a supply of white hydrophobic membrane filters of not more than 0.45 μm pore size and with high mechanical strength;

   c. a means of connecting the filter to the pipeline;

   d. a means of controlling the flow through the filter, which is connected downstream of the filter. One method of achieving this is to use the appropriate Amal jets to achieve a minimum flow of 150 L/min at 400 kPa and 350 L/min at 700 kPa;

   e. all equipment must be oxygen-compatible and hoses should be antistatic.

# Appendix E – Equipment for contaminant testing

## General

1. The function of these tests is to establish whether the pipeline has been contaminated during construction or modification. The specifications for the permissible concentrations of each component are summarised in Table 20.

2. Simple equipment that is of the required sensitivity and is suitable for use on site is commercially available.

## Measurement principle

3. A known volume of gas is passed through a tube packed with an absorbent, which is coated with specific colorimetric reagents. The reagents react quantitatively with the compound to be measured and produce a colour change along the length of the tube, which is proportional to the concentration of the compound being measured.

4. Tubes are available with appropriate sensitivities for the measurement of oil, water, carbon monoxide and carbon dioxide, sulphur dioxide, and higher oxides of nitrogen.

### Notes

Non-agent-specific detector tubes are difficult to interpret and are not recommended because of their qualitative and non-quantitative response.

# Appendix F – Equipment for gas identification

## General

1. The function of these tests is positively to identify medical gases by measuring their oxygen, nitrous oxide and nitric oxide content.

2. Portable equipment of the required specificity and sensitivity is commercially available.

3. Thermal conductivity meters do not give a positive identification of nitrous oxide in the presence of carbon dioxide, and should not be used as a sole means of identification of nitrous oxide. A specific nitrous oxide meter should be used. If carbon dioxide pipelines are present, for example in IVF (in vitro fertilisation) clinics, a carbon dioxide detector tube should be used.

## Specificity

### Oxygen

4. Oxygen-specific sensors using different measurement principles are currently in manufacture. The oxygen sensor should not give greater than ±1% response in the presence of 100% nitrous oxide or 100% nitrogen.

> **Notes**
>
> A paramagnetic meter is the specified instrument for identity of oxygen.

### Nitrous oxide

5. The nitrous oxide sensor should not give greater than ±1% response in the presence of 100% oxygen, 100% nitrogen or 100% carbon dioxide. An infrared/fuel cell meter is now commercially available.

## Specification

6. The equipment should be portable, preferably battery-powered, with digital or analogue indication of 0–100% to one decimal place. The battery should give at least eight hours' continuous running between recharging or replacement.

7. An accuracy better than ±1% is required, with a zero stability of 2.5% per day.

8. The response time must be not more than 15 seconds to 90% of the final reading.

# Appendix G – Pressure loss data

## Pipeline pressure-drop calculations

1. Example: Calculate the pressure drop in a 15 mm diameter pipe, 12 m in length, carrying medical air at a design flow rate of 800 L/min.

### Solution

2. The pressure drop $\Delta p$ across the pipe can be calculated from the formula:

$$\Delta p = \frac{\text{Measured length of pipe}}{\text{Nearest length of pipe from Table A1}} \times \left[\frac{\text{Design flow}}{\text{Nearest flow from Table A1}}\right]^2 \times \text{Pressure drop from Table A1} \quad (2)$$

3. From Table A1, the nearest length to 12 m is 15 m and the nearest flow rate to the design flow of 800 L/min is 711 L/min in the 15 m column, at which there is a pressure drop of 21 kPa across a 15 mm diameter, 15 m length of pipe.

4. Using these values, Equation (2) gives a pressure drop across the 12 m pipe of:

$$\frac{12}{15} \times \left[\frac{800}{711}\right]^2 \times 21$$

= 21.3 kPa.

5. If this loss is unacceptable, use the next (higher) pipe size, that is 22 mm. The nearest flow rate to 800 L/min is now 1135 L/min, representing a pressure loss of 7 kPa over 15 m.

6. In this instance:

$\Delta p$ = 2.8 kPa.

### Table A1 Section of pressure drop table for medical air

| Old BS 659 size | New British Standard size (BS EN 1057: R250, Table X) | | | | Distance from source at 400 kPa for 7, 14 and 21 kPa (0.07, 0.14 and 0.21 bar) pressure loss | |
|---|---|---|---|---|---|---|
| Nominal bore (inches) | Outside diameter (mm) | Wall thickness (mm) | Mean internal diameter (mm) | Mean internal diameter (inches) | 8 m (25 ft) | 15 m (50 ft) |
| 3/8 | 12 | 0.6 | 10.8 | 0.4252 | 311 455 564 | 209 307 382 |
| ½ | 15 | 0.7 | 13.6 | 0.5354 | 579 845 1038 | 391 572 711 |
| ¾ | 22 | 0.9 | 20.2 | 0.7953 | 1677 2441 3023 | 1135 1656 2053 |
| 1 | 28 | 1.2 | 26.2 | 1.0315 | 3363 4881 6034 | 2283 3320 4109 |
| ¼ | 35 | 1.2 | 32.6 | 1.2835 | 6023 8720 10,758 | 4096 5943 7344 |
| 1½ | 42 | 1.2 | 39.6 | 1.5591 | 10103 14,587 17,963 | 6883 9963 12,290 |

7. It is possible to insert the above formula into a spreadsheet and use mathematical functions to calculate required pressure drops (see Tables A2–A5).

## Table A2 Pipeline pressure loss: 400 kPa (4 bar) pipelines

| British Standard Size Tube BS EN 1057: R250, Table X | | Distance from source (m) at 400 kPa for 7, 14, 21 kPa (1, 2, 3 psi) pressure loss | | | | | | | | | | | | | | | | |
|---|---|---|---|---|---|---|---|---|---|---|---|---|---|---|---|---|---|---|
| Outside Diameter (mm) | Pressure loss (kPa) | 8 | 15 | 30 | 61 | 91 | 122 | 152 | 183 | 213 | 244 | 274 | 305 | 335 | 366 | 396 | 427 | 457 |
| | | Free air flow rate (L/min) | | | | | | | | | | | | | | | | |
| 12 | 7 | 311 | 209 | 141 | 95 | 75 | 64 | 56 | 50 | 46 | 43 | 40 | 37 | 35 | 34 | 32 | 31 | 30 |
| | 14 | 455 | 307 | 207 | 139 | 110 | 94 | 82 | 74 | 68 | 63 | 59 | 55 | 52 | 50 | 47 | 45 | 44 |
| | 21 | 564 | 382 | 258 | 174 | 138 | 117 | 103 | 93 | 85 | 78 | 73 | 69 | 65 | 62 | 59 | 57 | 55 |
| 15 | 7 | 579 | 391 | 263 | 177 | 140 | 119 | 105 | 94 | 86 | 80 | 75 | 70 | 66 | 63 | 60 | 58 | 56 |
| | 14 | 845 | 572 | 386 | 260 | 207 | 175 | 154 | 139 | 127 | 118 | 110 | 104 | 98 | 93 | 89 | 85 | 82 |
| | 21 | 1038 | 711 | 481 | 325 | 258 | 219 | 192 | 173 | 159 | 147 | 137 | 129 | 122 | 117 | 111 | 107 | 102 |
| 22 | 7 | 1677 | 1135 | 768 | 518 | 411 | 349 | 307 | 277 | 254 | 235 | 220 | 207 | 196 | 186 | 178 | 170 | 164 |
| | 14 | 2441 | 1656 | 1123 | 759 | 604 | 513 | 451 | 407 | 373 | 345 | 323 | 304 | 288 | 274 | 262 | 251 | 241 |
| | 21 | 3023 | 2053 | 1395 | 945 | 751 | 638 | 562 | 507 | 465 | 431 | 403 | 379 | 359 | 342 | 326 | 313 | 301 |
| 28 | 7 | 3363 | 2283 | 1547 | 1047 | 832 | 706 | 622 | 560 | 514 | 476 | 445 | 419 | 397 | 378 | 361 | 346 | 332 |
| | 14 | 4881 | 3320 | 2257 | 1530 | 1218 | 1035 | 912 | 823 | 754 | 699 | 653 | 615 | 583 | 555 | 530 | 508 | 488 |
| | 21 | 6034 | 4109 | 2800 | 1901 | 1514 | 1287 | 1135 | 1024 | 938 | 870 | 814 | 767 | 726 | 691 | 660 | 633 | 609 |
| 35 | 7 | 6023 | 4096 | 2783 | 1886 | 1500 | 1275 | 1124 | 1013 | 928 | 861 | 805 | 758 | 718 | 683 | 653 | 626 | 602 |
| | 14 | 8720 | 5943 | 4051 | 2752 | 2192 | 1865 | 1644 | 1483 | 1360 | 1261 | 1180 | 1111 | 1053 | 1002 | 957 | 918 | 883 |
| | 21 | 10758 | 7344 | 5018 | 3415 | 2723 | 2317 | 2044 | 1845 | 1692 | 1569 | 1468 | 1383 | 1310 | 1248 | 1192 | 1143 | 1099 |
| 42 | 7 | 10103 | 6883 | 4685 | 3180 | 2533 | 2154 | 1899 | 1713 | 1570 | 1456 | 1362 | 1283 | 1215 | 1157 | 1105 | 1060 | 1019 |
| | 14 | 14587 | 9963 | 6806 | 4633 | 3694 | 3145 | 2775 | 2504 | 2296 | 2130 | 1993 | 1878 | 1780 | 1694 | 1619 | 1553 | 1493 |
| | 21 | 17963 | 12290 | 8421 | 5743 | 4584 | 3904 | 3446 | 3112 | 2855 | 2648 | 2478 | 2335 | 2213 | 2107 | 2014 | 1932 | 1858 |
| 54 | 7 | 14974 | 10588 | 7487 | 5294 | 4323 | 3743 | 3348 | 3056 | 2830 | 2647 | 2496 | 2368 | 2257 | 2161 | 2076 | 2001 | 1933 |
| | 14 | 21176 | 14974 | 10588 | 7487 | 6113 | 5294 | 4735 | 4323 | 4002 | 3743 | 3529 | 3348 | 3192 | 3056 | 2937 | 2830 | 2734 |
| | 21 | 25935 | 18339 | 12968 | 9169 | 7487 | 6484 | 5799 | 5294 | 4901 | 4585 | 4323 | 4101 | 3910 | 3743 | 3597 | 3466 | 3348 |
| 76 | 7 | 37754 | 26696 | 18877 | 13348 | 10899 | 9438 | 8442 | 7706 | 7135 | 6674 | 6292 | 5969 | 5692 | 5449 | 5236 | 5045 | 4874 |
| | 14 | 53392 | 37754 | 26696 | 18877 | 15413 | 13348 | 11939 | 10899 | 10090 | 9438 | 8899 | 8442 | 8049 | 7706 | 7404 | 7135 | 6893 |
| | 21 | 65392 | 46239 | 32696 | 23119 | 18877 | 16348 | 14622 | 13348 | 12358 | 11560 | 10899 | 10339 | 9858 | 9438 | 9068 | 8738 | 8442 |

Examples:

1 122 m of 28 mm pipe would carry 706 L/min of free air per minute with a pressure loss of 0.07 bar (7 kPa), or 1287 L/min with a loss of 0.21 bar (21 kPa).
ie: $122/122 \times (706/706)^2 \times 7$

2 A flow of 1200 L/min in 122 m of 28 mm pipe would result in a pressure loss of 18.26 kPa. ie: $122/122 \times (1200/1287)^2 \times 21$

3 140 m of 28 mm pipe would carry 800 L/min with a pressure loss of 9.92 kPa. ie: $140/152 \times (800/912)^2 \times 14$

## Table A3 Pipeline pressure loss: 700 kPa (7 bar) pipelines

| British Standard Size Tube BS EN 1057: R250, Table X | | Distance from source (m) at 700 kPa for 7, 14, 34 kPa (1, 2, 5 psi) pressure loss | | | | | | | | | | | | | | | | |
|---|---|---|---|---|---|---|---|---|---|---|---|---|---|---|---|---|---|---|
| Outside Diameter (mm) | Pressure loss (kPa) | 8 | 15 | 30 | 61 | 91 | 122 | 152 | 183 | 213 | 244 | 274 | 305 | 335 | 366 | 396 | 427 | 457 |
| | | Free air flow rate (L/min) | | | | | | | | | | | | | | | | |
| 12 | 7 | 408 | 276 | 186 | 125 | 99 | 84 | 74 | 67 | 61 | 56 | 53 | 50 | 47 | 45 | 43 | 41 | 39 |
| | 14 | 599 | 405 | 274 | 185 | 147 | 124 | 109 | 99 | 90 | 84 | 78 | 74 | 70 | 66 | 63 | 61 | 58 |
| | 34 | 979 | 664 | 450 | 304 | 242 | 205 | 181 | 163 | 149 | 138 | 129 | 122 | 115 | 110 | 105 | 100 | 96 |
| 15 | 7 | 759 | 514 | 347 | 234 | 186 | 158 | 139 | 125 | 114 | 106 | 99 | 93 | 88 | 84 | 80 | 77 | 74 |
| | 14 | 1112 | 754 | 510 | 345 | 274 | 232 | 205 | 184 | 169 | 156 | 146 | 138 | 130 | 124 | 118 | 114 | 109 |
| | 34 | 1811 | 1231 | 836 | 566 | 450 | 383 | 337 | 304 | 279 | 258 | 242 | 227 | 215 | 205 | 196 | 188 | 180 |
| 22 | 7 | 2192 | 1488 | 1009 | 682 | 542 | 460 | 406 | 366 | 335 | 310 | 290 | 273 | 259 | 246 | 235 | 225 | 217 |
| | 14 | 3198 | 2175 | 1478 | 1001 | 797 | 677 | 597 | 538 | 493 | 457 | 428 | 403 | 381 | 363 | 347 | 332 | 320 |
| | 34 | 5180 | 3533 | 2410 | 1638 | 1306 | 1111 | 980 | 884 | 811 | 752 | 704 | 663 | 628 | 598 | 571 | 548 | 527 |
| 28 | 7 | 4387 | 2984 | 2027 | 1374 | 1093 | 929 | 819 | 739 | 677 | 628 | 587 | 553 | 524 | 498 | 476 | 456 | 439 |
| | 14 | 6382 | 4351 | 2963 | 2013 | 1604 | 1364 | 1203 | 1086 | 995 | 923 | 863 | 813 | 771 | 734 | 701 | 672 | 646 |
| | 34 | 10290 | 7038 | 4816 | 3283 | 2620 | 2232 | 1970 | 1779 | 1632 | 1514 | 1417 | 1335 | 1266 | 1205 | 1152 | 1105 | 1063 |
| 35 | 7 | 7841 | 5345 | 3638 | 2470 | 1968 | 1674 | 1476 | 1332 | 1221 | 1132 | 1059 | 998 | 945 | 900 | 860 | 825 | 793 |
| | 14 | 11380 | 7775 | 5307 | 3612 | 2881 | 2453 | 2165 | 1954 | 1792 | 1662 | 1556 | 1466 | 1389 | 1323 | 1264 | 1212 | 1166 |
| | 34 | 18271 | 12528 | 8599 | 5876 | 4696 | 4003 | 3536 | 3194 | 2931 | 2720 | 2547 | 2401 | 2276 | 2168 | 2073 | 1988 | 1912 |
| 42 | 7 | 13128 | 8964 | 6113 | 4159 | 3316 | 2823 | 2490 | 2248 | 2061 | 1912 | 1789 | 1686 | 1598 | 1521 | 1454 | 1394 | 1341 |
| | 14 | 19010 | 13012 | 8901 | 6070 | 4847 | 4129 | 3646 | 3293 | 3021 | 2803 | 2624 | 2473 | 2344 | 2232 | 2134 | 2047 | 1969 |
| | 34 | 30392 | 20892 | 14381 | 9849 | 7881 | 6723 | 5942 | 5371 | 4930 | 4577 | 4286 | 4042 | 3833 | 3651 | 3491 | 3349 | 3223 |

Examples:

1 122 m of 28 mm pipe would carry 929 L/min of free air per minute with a pressure loss of 0.07 bar (7 kPa), or 2232 L/min with a loss of 0.34 bar (34 kPa).
ie: $122/122 \times (929/929)^2 \times 7$

2 A flow of 1800 L/min in 122 m of 28 mm pipe would result in a pressure loss of 22.11 kPa. ie: $122/122 \times (1800/2232)^2 \times 34$

3 140 m of 28 mm pipe would carry 1100 L/min with a pressure loss of 10.78 kPa. ie: $140/152 \times (1100/1203)^2 \times 14$

## Table A4 Pipeline pressure loss: 1100 kPa (11 bar) pipelines

| British Standard Size Tube BS EN 1057: R250, Table X | | Distance from source (m) at 1100 kPa for 7, 14, 34 kPa (1, 2, 5 psi) pressure loss | | | | | | | | | | | | | | | | |
|---|---|---|---|---|---|---|---|---|---|---|---|---|---|---|---|---|---|---|
| Outside Diameter (mm) | Pressure loss (kPa) | 8 | 15 | 30 | 61 | 91 | 122 | 152 | 183 | 213 | 244 | 274 | 305 | 335 | 366 | 396 | 427 | 457 |
| | | Free air flow rate (L/min) | | | | | | | | | | | | | | | | |
| 12 | 7 | 487 | 356 | 252 | 177 | 144 | 124 | 112 | 102 | 94 | 88 | 84 | 79 | 75 | 72 | 69 | 67 | 65 |
| | 14 | 689 | 503 | 355 | 249 | 204 | 177 | 158 | 144 | 133 | 124 | 118 | 111 | 106 | 102 | 98 | 94 | 91 |
| | 34 | 1084 | 791 | 560 | 392 | 321 | 277 | 249 | 227 | 210 | 197 | 185 | 176 | 167 | 161 | 154 | 148 | 143 |
| 15 | 7 | 867 | 634 | 448 | 314 | 257 | 222 | 199 | 181 | 168 | 157 | 148 | 141 | 134 | 128 | 124 | 119 | 115 |
| | 14 | 1226 | 895 | 633 | 444 | 363 | 314 | 281 | 257 | 238 | 222 | 209 | 199 | 189 | 181 | 174 | 168 | 162 |
| | 34 | 1929 | 1409 | 996 | 698 | 572 | 494 | 443 | 403 | 373 | 350 | 330 | 313 | 298 | 285 | 275 | 264 | 256 |
| 22 | 7 | 2332 | 1703 | 1205 | 845 | 692 | 598 | 535 | 487 | 452 | 423 | 399 | 378 | 360 | 345 | 332 | 319 | 309 |
| | 14 | 3294 | 2405 | 1701 | 1193 | 977 | 844 | 755 | 689 | 638 | 597 | 562 | 534 | 509 | 487 | 468 | 451 | 436 |
| | 34 | 5185 | 3787 | 2678 | 1878 | 1537 | 1328 | 1189 | 1084 | 1005 | 939 | 886 | 840 | 801 | 767 | 737 | 710 | 686 |
| 28 | 7 | 4469 | 3263 | 2308 | 1618 | 1325 | 1145 | 1025 | 935 | 866 | 809 | 764 | 724 | 691 | 660 | 636 | 612 | 591 |
| | 14 | 6311 | 4608 | 3259 | 2286 | 1872 | 1616 | 1448 | 1320 | 1223 | 1143 | 1078 | 1022 | 976 | 933 | 897 | 864 | 835 |
| | 34 | 9935 | 7255 | 5130 | 3598 | 2946 | 2544 | 2279 | 2077 | 1926 | 1799 | 1698 | 1609 | 1535 | 1469 | 1412 | 1359 | 1315 |
| 35 | 7 | 7718 | 5636 | 3985 | 2795 | 2289 | 1976 | 1771 | 1614 | 1495 | 1397 | 1319 | 1250 | 1192 | 1141 | 1097 | 1056 | 1021 |
| | 14 | 10898 | 7959 | 5628 | 3947 | 3231 | 2791 | 2500 | 2279 | 2112 | 1973 | 1862 | 1765 | 1684 | 1611 | 1549 | 1492 | 1442 |
| | 34 | 17157 | 12530 | 8860 | 6213 | 5087 | 4394 | 3936 | 3587 | 3325 | 3107 | 2932 | 2779 | 2651 | 2537 | 2439 | 2348 | 2271 |
| 42 | 7 | 12550 | 9166 | 6481 | 4545 | 3721 | 3214 | 2879 | 2624 | 2432 | 2272 | 2144 | 2033 | 1940 | 1855 | 1784 | 1718 | 1661 |
| | 14 | 17724 | 12944 | 9152 | 6418 | 5255 | 4538 | 4066 | 3706 | 3435 | 3209 | 3029 | 2871 | 2739 | 2620 | 2519 | 2426 | 2345 |
| | 34 | 27902 | 20377 | 14409 | 10104 | 8273 | 7145 | 6401 | 5834 | 5407 | 5052 | 4768 | 4519 | 4312 | 4125 | 3966 | 3819 | 3692 |

Examples:

1 122 m of 28 mm pipe would carry 1145 L/min of free air per minute with a pressure loss of 0.07 bar (7 kPa), or 2544 L/min with a loss of 0.34 bar (34 kPa).
ie: $122/122 \times (1145/1145)^2 \times 7$

2 A flow of 2200 L/min in 122 m of 28 mm pipe would result in a pressure loss of 25.43 kPa. ie: $122/122 \times (2200/2544)^2 \times 34$

3 140 m of 28 mm pipe would carry 1300 L/min with a pressure loss of 10.39 kPa. ie: $140/152 \times (1300/1448)^2 \times 14$

## Table A5 Pipeline pressure loss (vacuum)

| British Standard Size Tube BS EN 1057: R250, Table X | | Distance from source (m) at 59 kPa (450 mm Hg) for 1.3, 2.6, 3.9, 6.5 kPa (10, 20, 30, 50 mm Hg) pressure loss | | | | | | | | | | | | | | | | |
|---|---|---|---|---|---|---|---|---|---|---|---|---|---|---|---|---|---|---|
| Outside Diameter (mm) | Pressure loss (kPa) | 8 | 15 | 30 | 61 | 91 | 122 | 152 | 183 | 213 | 244 | 274 | 305 | 335 | 366 | 396 | 427 | 457 |
| | | Free air flow rate (L/min) | | | | | | | | | | | | | | | | |
| 12 | 1.3 | – | – | – | – | – | – | – | – | – | – | – | – | – | – | – | – | – |
| | 2.6 | 47 | – | – | – | – | – | – | – | – | – | – | – | – | – | – | – | – |
| | 3.9 | 60 | 40 | – | – | – | – | – | – | – | – | – | – | – | – | – | – | – |
| | 6.5 | 82 | 55 | – | – | – | – | – | – | – | – | – | – | – | – | – | – | – |
| 15 | 1.3 | 59 | – | – | – | – | – | – | – | – | – | – | – | – | – | – | – | – |
| | 2.6 | 89 | 59 | 40 | – | – | – | – | – | – | – | – | – | – | – | – | – | – |
| | 3.9 | 113 | 76 | 51 | – | – | – | – | – | – | – | – | – | – | – | – | – | – |
| | 6.5 | 153 | 103 | 69 | 46 | – | – | – | – | – | – | – | – | – | – | – | – | – |
| 22 | 1.3 | 173 | 116 | 78 | 52 | 41 | – | – | – | – | – | – | – | – | – | – | – | – |
| | 2.6 | 260 | 174 | 117 | 79 | 62 | 53 | 46 | 42 | – | – | – | – | – | – | – | – | – |
| | 3.9 | 330 | 222 | 149 | 100 | 79 | 67 | 59 | 53 | 49 | 45 | 42 | 40 | – | – | – | – | – |
| | 6.5 | 445 | 301 | 203 | 137 | 108 | 92 | 81 | 73 | 67 | 62 | 57 | 54 | 51 | 49 | 46 | 45 | 43 |
| 28 | 1.3 | 350 | 236 | 159 | 106 | 84 | 71 | 63 | 56 | 51 | 48 | 44 | 42 | 40 | – | – | – | – |
| | 2.6 | 525 | 353 | 238 | 160 | 127 | 107 | 94 | 85 | 78 | 72 | 67 | 63 | 60 | 57 | 54 | 52 | 50 |
| | 3.9 | 666 | 448 | 303 | 204 | 161 | 137 | 120 | 108 | 99 | 92 | 86 | 81 | 76 | 73 | 69 | 66 | 64 |
| | 6.5 | 900 | 607 | 412 | 278 | 220 | 187 | 164 | 148 | 135 | 125 | 117 | 110 | 104 | 99 | 95 | 91 | 87 |
| 35 | 1.3 | 637 | 427 | 288 | 193 | 153 | 130 | 114 | 102 | 94 | 87 | 81 | 76 | 72 | 69 | 65 | 63 | 60 |
| | 2.6 | 947 | 638 | 431 | 290 | 230 | 195 | 171 | 154 | 141 | 131 | 122 | 115 | 109 | 103 | 99 | 95 | 91 |
| | 3.9 | 1198 | 808 | 548 | 369 | 293 | 248 | 218 | 197 | 180 | 167 | 156 | 147 | 139 | 132 | 126 | 121 | 116 |
| | 6.5 | 1614 | 1091 | 743 | 503 | 399 | 339 | 298 | 269 | 246 | 228 | 213 | 200 | 190 | 180 | 172 | 165 | 158 |
| 42 | 1.3 | 1074 | 724 | 488 | 328 | 260 | 220 | 194 | 174 | 160 | 148 | 138 | 130 | 123 | 117 | 111 | 107 | 103 |
| | 2.6 | 1598 | 1079 | 731 | 493 | 391 | 331 | 291 | 262 | 240 | 222 | 208 | 196 | 185 | 176 | 168 | 161 | 155 |
| | 3.9 | 2016 | 1363 | 926 | 626 | 497 | 422 | 371 | 334 | 306 | 283 | 265 | 249 | 236 | 224 | 214 | 205 | 197 |
| | 6.5 | 2706 | 1833 | 1254 | 851 | 677 | 574 | 506 | 456 | 417 | 387 | 361 | 340 | 322 | 306 | 293 | 280 | 270 |
| 54 | 1.3 | 2191 | 1480 | 1001 | 674 | 535 | 453 | 399 | 359 | 329 | 304 | 284 | 268 | 253 | 241 | 230 | 220 | 212 |
| | 2.6 | 3246 | 2196 | 1493 | 1010 | 802 | 681 | 599 | 540 | 494 | 458 | 428 | 403 | 381 | 363 | 346 | 332 | 319 |
| | 3.9 | 4083 | 2766 | 1889 | 1281 | 1019 | 865 | 762 | 687 | 629 | 582 | 545 | 513 | 485 | 462 | 441 | 423 | 406 |
| | 6.5 | 5448 | 3699 | 2549 | 1737 | 1384 | 1176 | 1037 | 935 | 856 | 794 | 742 | 699 | 662 | 630 | 601 | 576 | 554 |
| 76 | 1.3 | 5521 | 3773 | 2563 | 1733 | 1377 | 1169 | 1029 | 927 | 849 | 786 | 735 | 692 | 655 | 623 | 595 | 570 | 548 |
| | 2.6 | 8070 | 5563 | 3807 | 2586 | 2058 | 1749 | 1541 | 1389 | 1273 | 1179 | 1103 | 1038 | 983 | 936 | 894 | 857 | 823 |
| | 3.9 | 10041 | 6968 | 4801 | 3274 | 2609 | 2219 | 1957 | 1765 | 1617 | 1499 | 1402 | 1320 | 1250 | 1190 | 1137 | 1090 | 1048 |
| | 6.5 | 13166 | 9233 | 6439 | 4421 | 3533 | 3009 | 2655 | 2396 | 2197 | 2037 | 1906 | 1796 | 1701 | 1619 | 1547 | 1483 | 1426 |
| 108 | 1.3 | 12874 | 9140 | 6543 | 4552 | 3732 | 3280 | 2879 | 2628 | 2433 | 2276 | 1919 | 2036 | 1941 | 1858 | 1785 | 1712 | 1641 |
| | 2.6 | 18207 | 12874 | 9235 | 6578 | 5274 | 4552 | 4071 | 3716 | 3441 | 3219 | 3035 | 2879 | 2745 | 2628 | 2525 | 2422 | 2325 |
| | 3.9 | 22494 | 15905 | 11374 | 8114 | 6509 | 5657 | 5030 | 4592 | 4251 | 3976 | 3750 | 3557 | 3391 | 3247 | 3119 | 2992 | 2870 |
| | 6.5 | 29238 | 20675 | 14708 | 10520 | 8445 | 7343 | 6538 | 5968 | 5526 | 5169 | 4873 | 4623 | 4408 | 4220 | 4055 | 3889 | 3730 |

Examples:

1 122 m of 28 mm pipe would carry 71 L/min of free air per minute with a pressure loss of 10 mm Hg (1.3 kPa), or 187 L/min with a loss of 50 mm Hg (6.5 kPa).
ie: $122/122 \times (71/71)^2 \times 1.3$

2 A flow of 120 L/min in 122 m of 28 mm pipe would result in pressure loss of 2.99 kPa. ie: $122/122 \times (120/137)^2 \times 3.9$

3 140 m of 28 mm pipe would carry 90 L/min with a pressure loss of 2.20 kPa. ie: $140/152 \times (90/94)^2 \times 2.6$

8. Another alternative is to derive graphs from the tables, although it may be necessary to draw several graphs, at different scales, to obtain accurate results.

9. The graphs of flow vs pressure drop provide a pressure loss per metre of pipe, not a total pressure loss. This figure must be multiplied by the length of the pipe in order to find the actual total pressure drop.

10. Because a pipe and the fittings in the system cause frictional resistance to the gas flow, a pressure loss occurs that is greater than that which would occur if the gas were flowing through the same distance of straight pipe.

11. Each valve, fitting etc is allocated a "length" equivalent in frictional resistance to a straight piece of pipe of the same diameter. This length is hence known as the equivalent length of the fitting.

12. To calculate design flows, the sum of the lengths of the straight runs of pipe plus the sums of the equivalent lengths of all of the fittings etc in that run are added.

13. In practice many designers simply add 25–30% to the total measured length or use only 60–75% of the allocated pressure drop when sizing.

14. Equivalent lengths of some fittings are given in Tables A6 and A7.

## Table A6 Equivalent lengths for copper fittings

| | 6 mm | 8 mm | 10 mm | 12 mm | 15 mm | 22 mm | 28 mm | 35 mm | 42 mm | 54 mm | 76 mm |
|---|---|---|---|---|---|---|---|---|---|---|---|
| Ball valve | 0.10 | 0.10 | 0.20 | 0.30 | 0.30 | 0.60 | 0.90 | 0.90 | 1.10 | 1.20 | 1.20 |
| Tee (Thru') | 0.12 | 0.15 | 0.18 | 0.21 | 0.32 | 0.42 | 0.54 | 0.70 | 0.82 | 1.05 | 1.56 |
| Tee (Branch) | 0.46 | 0.52 | 0.70 | 0.80 | 0.95 | 1.26 | 1.60 | 2.10 | 2.45 | 3.14 | 4.67 |
| 90° Elbow | 0.17 | 0.20 | 0.25 | 0.33 | 0.47 | 0.63 | 0.80 | 1.05 | 1.23 | 1.58 | 2.36 |

## Table A7 Equivalent lengths for ABS (acrylonitrile butadiene styrene) vacuum fittings

| | 40 mm | 50 mm | 70 mm | 100 mm | 125 mm |
|---|---|---|---|---|---|
| Tee (Thru') | 0.95 | 1.23 | 1.65 | 2.20 | 2.56 |
| Tee (Branch) | 2.76 | 3.38 | 4.57 | 6.12 | 7.68 |
| 90° elbow | 1.25 | 1.71 | 2.44 | 3.08 | 3.84 |

# Appendix H – Checklist for planning/installing/upgrading a cryogenic liquid supply system

1. Information given in this Appendix can be used to determine the need for a particular capacity or type of supply system. Many of the factors described will also apply to planning an upgrade to an installation by way of increase in system size or a change of system type.
2. Some factors that should be considered are outlined below.

## Delivery frequency

- Does current frequency cause logistical problems for the supplier/your site?

## Calculating consumption

- Consumption is rising at approximately 10% per annum. It doubles in seven years.
- Use pharmacy records for cylinder/liquid consumption. Look for peaks in demand, for example winter influenza epidemics.
- When average and peak flow rates are known, calculate the required size of the emergency supply.

## Age of current system

- The secondary supply of older VIE systems will be a compressed gas cylinder manifold, which may have very limited capacity. Consideration should be given to either a single VIE plus fully automatic manifold or, preferably, a dual VIE system.

## Siting of system and the site survey

- What planning restrictions apply (vessel size, noise etc)?
- What are convenient locations for cylinder/liquid delivery?
- Advantages of separating primary and secondary supplies, if space is available.
- Will other facilities be lost/reduced, for example car-parking space?
- It will be less economical in terms of delivery charges and unit gas costs to deliver large loads (for example 20 tons) using rigid vehicles (maximum 12 tons). Articulated vehicles will deliver the largest loads but may require roadway/access modifications.
- Cranage access for vessels.
- When choosing liquid cylinder systems, will adequate ventilation be available?
- Emergency supply location.
- Pipeline protection and possible need for dual feeds.
- Pipeline extension into other sites if applicable, for example two hospitals supplied from the same VIE system. There are possible insurance issues with this arrangement.
- Modifications to the alarm system may have to be made.
- Alarm panel + telemetry in waterproof enclosures.
- Are alarms compatible with the existing system?
- Alarm arrangement for dual (but separate) tank installations.
- Cable ducts and trays: examine possible routes.
- Possible need to move gates/fences to install new pipework.
- Clearance of trees/building.
- Sealing windows of adjacent buildings.
- Position of frame for valve tree (fix to fence for rigidity?).
- Position of emergency gate.
- Position of fill couplings must allow driver to see tank gauges.

- Cabling and alarm runs for the emergency supply manifold (ESM).
- Availability and presentation of alarms for ESM.
- Power and lighting during work.
- Drainage – catch pits, diversions, pad resizing.

## Costs

- Make sure **all** costs are allowed for, for example:
  - (i) Site inspection.
  - (ii) Cost of continuing delivery using rigid and non-articulated vehicles.
  - (iii) Gas charge/HCM (hundred cubic metres) and any inflation likely.
  - (iv) Facility charges (rental).
  - (v) Delivery charge for equipment.
  - (vi) Loan charges and changes in interest rate on any loan if the installer funds any part of the installation.
  - (vii) Road/compound loans will be seen as £*x* added to gas price over *y* years.
  - (viii) Climate change levy.
  - (ix) Professional fees (consultancy).
  - (x) Planning permission.
  - (xi) Building Regulations clearance.
  - (xii) All civil engineering work.
  - (xiii) Quoted price for gas/facilities/delivery charges may be dependent on payment by direct debit.
  - (xiv) Introduction/modification and maintenance of services, for example lighting, power supplies, drainage.
  - (xv) Engineering and pharmaceutical testing.
  - (xvi) Additional emergency provision and any associated cylinder charges.
  - (xvii) Modifications to alarm and telephone systems.
  - (xviii) Security.
  - (xix) Charges for ESM cylinders during installation (may have to be charged and then recovered).
  - (xx) Cranage charges.
  - (xxi) Contingency 10%.
- What, if any, commitment is required by the gas company?
- How will gas prices vary during this period?
- Is there any agreement to provide, for example, modified roadway facilities if rigid vehicular deliveries are too frequent to be convenient to supplier? Or if such roadway modifications take place within a defined timescale, new rates etc may need to be negotiated.
- Check defects liability (usually 12 months).

## Emergency provision

- Examine the vulnerability of current system and main feeds to hospital.
- Consider minimum size of manifold plus cylinder storage to meet four-hour supply requirement. Is a second VIE a better option?
- Operational requirements of ESM.
- Protection/housing/security of ESM.
- Alarm/monitoring systems and power supplies for ESM and its accommodation.

## System shutdown during installation

- Often it will be necessary to interrupt site supplies during connection of new plant. How will this be managed?
- Disruption of two hospitals simultaneously if plant to be upgraded is supplying both sites.
- Examine planned plant and pipework systems carefully to ascertain the best way of minimising downtime and facilitating engineering and pharmaceutical testing.
- While installing, fit extra valves to allow for future expansion and emergency supply manifolds to protect vulnerable parts of the system.
- Fit NIST fittings wherever this will facilitate system purging.
- Fit test points/emergency inlet ports as recommended in this guidance or investigate any likely requirement for additional (local) manifolds to support high-dependency areas.

## Paperwork

- Site survey details.
- Register of contractors with contact names/ telephone numbers.
- Keep a record of all dates, for example:
  - (i) tender invitation;
  - (ii) tender open;
  - (iii) tender close;
  - (iv) award and regret letters to tenderers.
- Copies of all letters to/from contractors.
- NICEIC (National Inspection Council for Electrical Installation Contracting) test certificate for electrics.
- Validation and verification results (engineering and pharmaceutical).
- Health and safety policies of contractors.
- Method statements from contractors.
- Insurance agreement with gas supplier for VIE system(s).
- MGPS operational policy protocols.

## Health and safety

- Health and safety policy (contractors and their employees, and subcontractors and their employees, must comply when employed by the trust and working on trust properties).
- Inform contractors of specific site hazards.
- Hazard notices on site and on final installation.
- Lighting during installation and for completed compound.
- Road markings and signage.

## Preparation

- Carefully plan phasing of building work to maximise efficiency of installation programme. (Remember concrete plinths will take three days to harden before vessels can be sited.)
- Plan phasing of engineering and QC testing to avoid wasting APs'/QCs' time.
- Consider methods of maintaining supplies during essential shutdowns. Cylinder supplies may be needed during commissioning. Gas supplier may be able to arrange multi-cylinder pallets.
- Road base preparation, if required, must be completed in an early phase of the work to allow necessary access for cranes and, eventually, delivery vehicles.
- Road surfacing/kerbing/drainage/lighting.
- Retaining walls around compound if required, for example on sloping sites.
- Maintaining rights of way.
- Oxygen compound civil engineering work.
- If you are changing supplier, your original supplier will need to remove old equipment before plinth can be extended to fit new vessels.
- Electrics for alarms, tank, lighting and, possibly, vehicle pump.
- Floodlighting and telephone line.
- Plan vehicular parking during (and after) work.
- The old plinth may require skimming to provide a reasonable surface.

## Installation

- If an ESM (as a third means of supply) is installed first, this can be used to supply the hospital system during vessel replacement.
- Decide who arranges emergency cylinder supplies for ESM. When plinth extensions are required, specify oxygen-compatible sealant for gaps between old and new plinth sections.
- Remember to post health and safety notices during the work.
- Alarm systems will not be fully functional until system is fully commissioned. Therefore, all staff must be kept aware of the different alarm situation.
- Concrete will need two to three days to harden on any pad extension.
- The first vessel filling is a very noisy procedure with much vapour and can take several hours (consider restrictions).
- Concrete sample testing will be required during new plinth construction.
- Use temporary steel sheeting to support a new vessel on tarmac alongside the plinth.

- Access for cranage must be kept open (car parking control).
- Drainage (may have to move existing drains/ soakaways and create new pipe runs; remember oxygen separation distances).
- Road markings and signage.
- Possible new kerbs/footpaths.
- Electrical supplies: single phase can be used for lighting, alarms etc but three-phase 60 A supply will be needed for delivery vehicle pump if appropriate.
- Earth bonding/lightning protection for fences.
- Alarm interface/telemetry boxes at a sensible height for viewing.
- Lagging of liquid lines.
- If using 200 bar unregulated cylinders for supply during installation or on ESM, take care that they are not mixed up with 137 bar cylinders.
- Proximity of flammables and vital services during installation – vulnerability to mechanical damage (cutting discs etc), welding and cutting flames/sparks.
- Power and lighting supplies during work.
- Water supply (washing and concrete) during work.

## Follow-up

- Routine maintenance and monitoring of complete installation.
- Cylinder changes and stock management for ESM.
- Establish system management arrangements for vessels supplying more than one site (see Part B, Appendix G).
- Update MGPS operational policy and any relevant insurance policies.

# Appendix J – Upgrading surgical air systems

## Background

1. An increasing number of surgical air pipeline systems are designed to operate at a line pressure above the nominal 700 kPa, for example 1000–1100 kPa. This enables the system to deliver 350 L/min (at 700 kPa) at the front of the surgical air terminal unit.

2. Such a system will comprise a high-pressure supply pipeline installation, in the order of 1000–1100 kPa and local pressure regulation (for example adjacent to the operating suite) such that the maximum static pressure does not exceed 900 kPa.

3. Existing Health Technical Memorandum 2022 systems run at a line pressure of 700 kPa and will provide a flow of 250 L/min at the front of the terminal unit.

4. Users must be made aware (preferably by written report) that such systems will not meet the demands of some modern air tools and that use of such tools may result in both a lack of tool performance and frequent low-pressure alarms on the surgical air system.

5. Tools are available that require a flow up to 500 L/min at an operating pressure of 1400 kPa. Such tools will require discrete cylinder supplies.

## Modifying "old" systems

6. Increasing line pressure to meet the latest recommended flow rates is often proposed, but needs to take account of the following:

   a. **Is the compressor receiver suitable for use at the proposed pressures?**
   A typical "old" 700 kPa system will employ a receiver operating at typically 10 bar pressure. A "new" system, with a typical line pressure of 10.5 bar, requires a receiver operating at a typical pressure of 13 bar.
   Ensure that the test, design and working pressures of the current receiver are acceptable, and that the capacity of the receiver is appropriate to the new demand.

   b. **Is the compressor plant capable of meeting the increased duty cycle?**
   Overheating and premature plant failure may result if this is not the case. The system may be supplying both surgical and medical air. Plant failure/flow reduction resulting from an overburdened surgical air system may have serious consequences in terms of medical air provision, particularly as this is the recommended driving gas for patients' ventilators.
   This guidance recommends the use of separate plant for surgical air/medical air, but this may not always be the case with older Health Technical Memorandum 2022 plant.

   c. **Are the pressure safety valves (PSVs) suitably rated?**
   PSVs on pipelines and the receiver will need to be changed to meet the new operating conditions. Certificated replacement PSVs should be used.

   d. **Are pressure switches suitably rated and adjusted?**
   Pressure switches on plant and pipelines will need to be changed or adjusted accordingly.

   e. **Has the pipeline been suitably pressure-tested?**
   There may be occasions when existing 4 bar systems (or parts thereof) are proposed for use at 7 bar or higher. 4 bar systems are only tested to 10 bar; they will need to be retested at 18 bar to ensure a leak-free high-pressure system.

   f. **Will you still be insured?**
   If the pipeline system contains large diameter pipe (120 mm or above), the insurance company should be consulted to ascertain whether the system would be capable of withstanding not only the pipeline operating pressure but also any test pressure that may be applied during system refurbishment or extension. Many insurance companies will not allow testing of pipe diameters greater than

100 mm at 18 bar, as this pressure is likely to compromise the pipe integrity.
The insurance company should be consulted on any necessary amendments.
A new Written Scheme of Examination will have to be prepared for the new system.

g. **Labelling**.
Ensure that **all** relevant labels are in place before the system is accepted.

h. **Other issues.**
There may be other system defects discovered during the upgrading process, for example lack of pipeline support/protection.
There may also be a potential contamination issue resulting from the transfer of particulate matter from older, silver-soldered systems into new inert gas shield brazed pipework, although it should be noted that this is not an issue limited to air systems. The Quality Controller (MGPS) will advise on this and adopt appropriate testing methods.
These issues will need to be addressed before the system is accepted for use.
Ensure that any system amendments and changes in working practices are documented in the MGPS operational policy.

# Appendix K – Signage requirements

| Location | Wording | Notes |
|---|---|---|
| Plantroom | Medical Gas Plant Room – No unauthorised Entry | Adjacent to or on external door |
| | Fire action | On door/wall<br>External or internal |
| | Keep locked | On door(s) |
| | Noise Hazard (+ ear defender symbol)<br>Electric shock hazard<br>Permit-to-work must be used | Adjacent to, or on, external door |
| | Plant is connected to essential electricity supply | "E" symbols can be used on switches etc |
| | Danger 415 volts | On plant/switchgear |
| | Danger 240 volts | On plant/switchgear |
| | Danger rotating machines<br>Warning: These machines stop and start automatically without warning<br>Guards must be in position<br>Do not isolate without a Permit | Posted adjacent to plant |
| | Biological symbol | Vac filters and exhausts<br>Also for AGSS units/exhausts/drain flasks |
| | Medical air intake Do not obstruct | On external intakes only |
| | Emergency Tel No<br>Gas Supplier<br>Estates<br>Pharmacy<br>Porters | External wall |
| | Health and Safety Law | Internal wall |
| | First aid | Internal wall |

**Notes:**

"Bacteria filter change procedure" sign is not available commercially and will have to be made locally.

No "Danger medical gas/vac/AGSS exhaust" sign is commercially available but "Danger explosive gases, no smoking, no naked lights" is available and would suffice.

"Danger 440/240 volts", "Warning: These machines stop and start automatically/without warning" and "Biohazard" labels would need to be added to AGSS plant remote from main plantroom, plus any relevant plantroom notices.

| Location | Wording | Notes |
|---|---|---|
| Manifold room | Medical gases manifold room – No unauthorised entry | Adjacent to or on external door |
| | No parking | Adjacent to or on external door |
| | Approved personal protective equipment must be worn | Adjacent to or on external door |
| | Fire action | Internal/external wall |
| | Cylinder status tag | On manifold cylinders |
| | Valve open | On line valves/ERM cyls |
| | Valve closed | On line valves/ERM cyls |
| | Make sure cylinders are secure at all times | Internal, near cylinders |
| | Danger No smoking | External (on door or wall) |
| | Danger compressed gas | External (on door or wall) |
| | Warning oxidising agent | External (on door or wall) |
| | Danger oxygen | External (on door or wall) |
| | Emergency Tel No<br>Gas suppliers<br>Estates<br>Pharmacy<br>Porters | External, on wall or door |
| | Keep locked | On door |

**Notes:**

Also required:

Cylinder ID charts, manifold cylinder change procedure, emergency manifold operating procedure.

Check with fire officer for any local fire brigade requirements for fitting "HAZCHEM" signs eg "HAZCHEM 2SE Cylinders".

| Location | Wording | Notes |
|---|---|---|
| **Main cylinder store** | Medical gases cylinder store – No unauthorised entry | Adjacent to or on external door |
| | No parking | Adjacent to or on external door |
| | Keep loading bay/doors clear | Adjacent to or on external door |
| | Approved personal protective equipment must be worn | Adjacent to or on external door |
| | Make sure cylinders are secure at all times | Adjacent to cylinders |
| | Fire action | Internal/external wall |
| | Full cylinders | On bays |
| | Empty cylinders | On bays |
| | Emergency exit keep clear | May be already fitted |
| | Danger No smoking | On door |
| | Danger compressed gas | On door |
| | Warning oxidising agent | External (on door or wall) |
| | Danger oxygen | External (on door or wall) |
| | Emergency Tel No<br>Gas suppliers<br>Estates<br>Pharmacy<br>Porters | External wall |
| | Keep locked | On door |
| | Push bar to open | Emergency exit and main door(s) |

**Notes:**

"Danger liquid nitrogen" sign is available for a separate liquid nitrogen store (see BCGA CP30).

Cylinder ID chart(s) to be posted

Check with fire officer for any local fire brigade requirements for fitting "HAZCHEM" signs eg "HAZCHEM 2SE Cylinders".

| Location | Wording | Note |
|---|---|---|
| **Ready to use cylinder store** | Medical gases cylinder store – No unauthorised entry | Adjacent to or on external door |
| | Make sure cylinders are secure at all times | Adjacent to cylinders |
| | Emergency Tel No<br>Gas suppliers<br>Estates<br>Pharmacy<br>Porters | External wall |
| | Danger No smoking | On door |
| | Danger compressed gas | On door |
| | Fire action | Internal/external wall |

Notes:

Post cylinder ID chart(s) and cylinder change procedure.

Check with fire officer for any local fire brigade requirements for fitting "HAZCHEM" signs, eg "HAZCHEM 2SE Cylinders"

| Location | Wording | Notes |
|---|---|---|
| **Ward (cylinder parking bay)** | Medical gas cylinder parking area | Defines cylinder parking as per new Health Technical Memorandum |
| | Emergency Tel No<br>Gas suppliers<br>Estates<br>Pharmacy<br>Porters | External wall |
| | Gas leak action | On nurses' station |

Notes:

Post cylinder chart(s) and cylinder change procedure.

Operational policy may dictate posting of AVSU emergency operation and MGPS alarm responses.

| Location | Wording | Notes |
|---|---|---|
| **Work area** | Maintenance in progress | There may be other site safety notice requirements to fulfil |
| | Medical gas test area | |
| | Confined space | |
| | Hot work in progress | |
| | Danger pressure test in progress | |
| | Danger nitrogen purging in progress | |

Notes:

These signs should be posted during installation/modification/maintenance of an MGPS.

Multiple signs may be required.

| Location | Wording | Notes |
|---|---|---|
| **Pipework** | Gas identity | |
| | Flow direction | |

| Location | Wording | Note |
|---|---|---|
| **Line valves** | Gas identity | On pipeline label |
| | Flow direction | |
| | Key No | |

| Location | Wording | Notes |
|---|---|---|
| **AVSUs** | In emergency break glass and shut off valve | On/off positions to be shown on AVSU body |
| | Gas identity | |
| | Flow direction | |
| | Area controlled | |
| | Key No | |
| | Valve No | |

| Location | Wording | Notes |
|---|---|---|
| **Alarm displays** | Area monitored | Responses may be posted nearby, in accordance with MGPS operational policy |
| | Gas names | |
| | Fault/normal/condition indicators | |

| Location | |
|---|---|
| **VIE/ Liquid cylinders/PSA/synthetic air** | Signage will be determined by the equipment supplier but will usually include plant schematic, safety warnings and emergency actions |

# Appendix L – Important notes for use of medical vacuum and anaesthetic gas scavenging

## Infectious disease units

1. Medical vacuum should neither be extended to an Infectious Diseases Unit (IDU) nor provided to such a unit from a central vacuum system.

2. Portable suction units will be required. Decontamination will require specialised protocols and the advice of the infection control officer should be sought.

### Note

Systems already exist whereby an IDU is, by local agreement, serviced via a central vacuum system.

If such agreements exist, or are to be accepted, great care must be taken to ensure that the exhausts of such a plant are kept well away from all air intakes and the plant is labelled to indicate its function. Ideally, the plant should be housed in separate accommodation but, where this is not possible, safety signage and strict operational protocols are extremely important.

Personnel changing filters, or carrying out work on such a system, should wear personal protective equipment and follow protocols that have been devised in liaison with the infection control officer.

## Laser/surgical diathermy smoke extraction

3. An additional contamination hazard can arise if smoke from procedures employing laser or surgical diathermy equipment is exhausted using a cannula attached to the vacuum system.

4. Clinical staff should be advised against this practice and either instructed to use dedicated laser smoke removal units (incorporating dedicated, filtered, portable vacuum pumps) or a specially designed laser smoke filter fitted to a medical vacuum system terminal unit.

## Dental vacuum systems

5. Medical vacuum systems operate at relatively low flow rates at the terminal units (~40 L/min). Such flows are unsuitable for use as dental vacuum, which operates at much higher flow rates (typically 300 L/min). Medical vacuum systems should not be used to provide dental vacuum.

## Anaesthetic gas scavenging (AGS)

### Active AGS systems (medical)

6. Active anaesthetic gas scavenging systems operate at relatively high flow rates (80–130 L/min in a BS 6834:1987 system; and 50–80 L/min in an ISO/DIS 7396-2:2005 system).

7. It is unlikely that receiving systems designed for use with these scavenging systems will operate correctly with a medical vacuum system. Severe spillage of waste gases into the operating area may occur. Therefore, medical vacuum systems should not be used as waste anaesthetic gas disposal systems.

### Active AGS (dental)

8. Active AGS systems for use with dental nasal scavenging masks operate by maintaining a flow of air through the outer layer of a specially designed concentric nose mask. Waste gases from the patient pass from inner to outer layers of the mask and are carried away to the exhaust termination by this air stream.

9. The flow rate necessary to achieve effective removal of waste gases by such a method is in the order of 45 L/min, which is less than the flow rate achieved at a medical vacuum system terminal.

10. Active dental scavenging systems using this type of mask must therefore be driven (via a special flow adjuster) from the dental vacuum system, a dedicated separate high-flow vacuum system, or an active medical AGS system. In the case of a medical AGS system, the special flow adjuster would be plugged directly into an AGS system wall terminal. A receiver (air break) system would not be used between the wall terminal and the special flow adjuster.

# Appendix M – Oxygen usage data

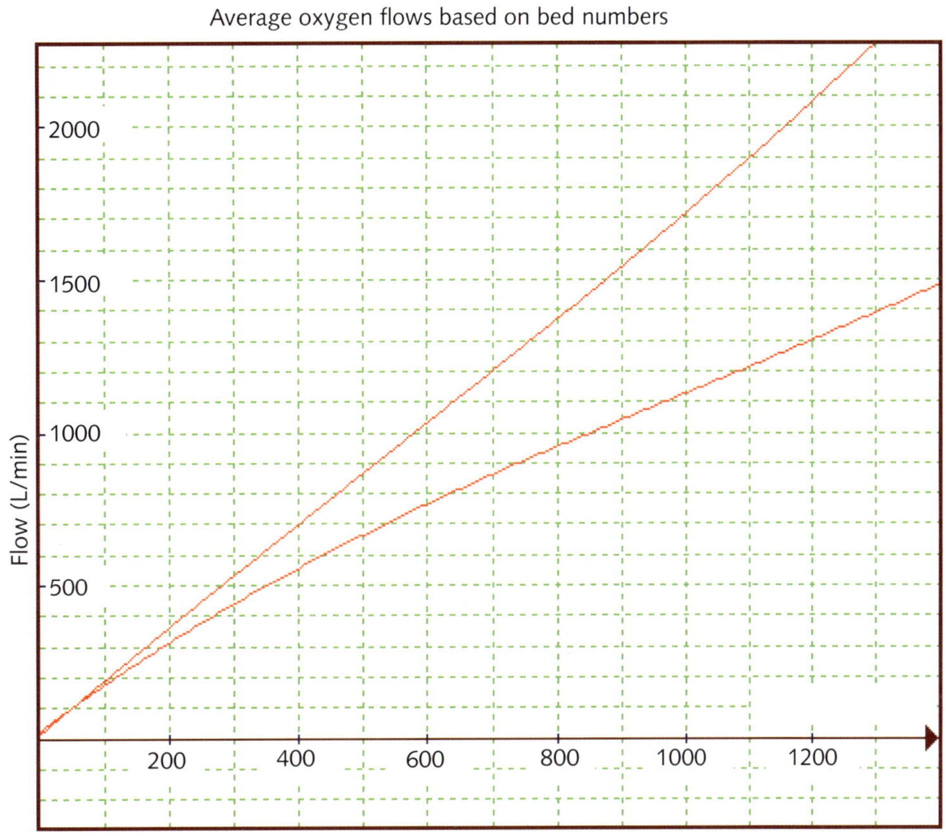


**Lower graph** shows average flows to be expected in a typical acute hospital.

**Upper graph** shows flows to be expected when the hospital has specialties using larger amounts of oxygen, eg those with multiple large critical care areas (>20 beds) and an increased use of CPAP (>5 machines).

NB These graphs are issued for guidance only. There will be hospitals for which average flows will, for a given number of beds, be higher or lower than the maxima and minima shown here.

NB The flows are representative of oxygen provided from a VIE plant and do NOT take into account additional consumption from compressed gas cylinders.

# Appendix N – Pressure conversion table

| Pressure | Multiply units in left column by factor below | | | | | | | |
|---|---|---|---|---|---|---|---|---|
| | kPa | $lb/in^2$ | $lb/ft^2$ | int atm | $kg/cm^2$ | mm Hg @0°C | in Hg @0°C | ft water @4°C |
| 1 pound/sq in | 6.895 | 1 | 144 | 0.0682 | 0.0703 | 51.713 | 2.0359 | 2.307 |
| 1 pound/sq ft | 0.048 | 0.00694 | 1 | 0.0005 | 0.00052 | 0.3591 | 0.01414 | 0.01602 |
| 1 int atmosphere | 101.3 | 14.696 | 2116.2 | 1 | 1.0333 | 760 | 29.921 | 33.9 |
| 1 kilogram/sq cm | 98.07 | 14.223 | 2048.1 | 0.9678 | 1 | 735.56 | 28.958 | 32.81 |
| 1 mm Hg (1 torr) | 0.133 | 0.0193 | 2.785 | 0.0013 | 0.00136 | 1 | 0.0394 | 0.0446 |
| 1 in Hg | 3.387 | 0.4912 | 70.73 | 0.0334 | 0.0345 | 25.400 | 1 | 1.133 |
| 1 ft water | 2.984 | 0.4335 | 62.42 | 0.0295 | 0.0305 | 22.418 | 0.8826 | 1 |
| 1 kilopascal (kPa) | 1 | 0.145 | 20.92 | 0.0099 | 0.0102 | 7.519 | 0.295 | 0.3346 |

# References


## Acts and Regulations

**(The) Control of Substances Hazardous to Health Regulations 2002.** SI 2002 No. 2677. HMSO, 2002. http://www.opsi.gov.uk/si/si2002/20022677.htm

**(The) Electromagnetic Compatibility Regulations 2005.** SI 2005 No. 281. HMSO, 2005. http://www.opsi.gov.uk/si/si2005/20050281.htm

**(The) Medicines Act 1968.** HMSO, 1968.

**(The) Pressure Systems and Transportable Gas Containers Regulations 1989.** SI 1989 No. 2169. HMSO, 1989. http://www.opsi.gov.uk/si/si1989/Uksi_19892169_en_1.htm

**(The) Public Health Act 1961.** HMSO, 1961.

**(The) Public Health (Drainage of Trade Premises) Act 1937.** HMSO, 1937.

**(The) Water Industry Act 1991.** HMSO, 1991.

**(The) Water Resources Act 1991.** HMSO, 1991.

## British Standards

**BS 88.** Cartridge fuses for voltages up to and including 1000 V ac and 1500 V dc.

**BS 341-3:2002.** Transportable gas container valves. Valve outlet connections. British Standards Institution, 2002.

**BS 476-4:1970.** Fire tests on building materials and structures. Non-combustibility test for materials. British Standards Institution, 1970.

**BS 476-20:1987.** Fire tests on building materials and structures. Method for determination of the fire resistance of elements of construction (general principles). British Standards Institution, 1987.

**BS 476-21:1987.** Fire tests on building materials and structures. Methods for determination of the fire resistance of loadbearing elements of construction. British Standards Institution, 1987.

**BS 476-22:1987.** Fire tests on building materials and structures. Methods for determination of the fire resistance of non-loadbearing elements of construction. British Standards Institution, 1987.

**BS 476-23:1987.** Fire tests on building materials and structures. Methods for determination of the contribution of components to the fire resistance of a structure. British Standards Institution, 1987.

**BS 1710:1984.** Specification for identification of pipelines and services. British Standards Institution, 1984.

**BS 3928:1969.** Method for sodium flame test for air filters (other than for air supply to IC engines and compressors). British Standards Institution, 1969.

**BS 5499-5:2002.** Graphical symbols and signs. Safety signs, including fire safety signs. Signs with specific safety meanings. British Standards Institution, 2002.

**BS 5682:1998.** Specification for probes (quick connectors) for use with medical gas pipeline systems. British Standards Institution, 1998.

**BS 6834:1987.** Specification for active anaesthetic gas scavenging systems. British Standards Institution, 1987.

**BS 7671:2001.** Requirements for electrical installations. IEE Wiring Regulations. Sixteenth edition. British Standards Institution, 2001.

**BS EN 286-1:1998.** Simple unfired pressure vessels designed to contain air or nitrogen. Pressure vessels for general purposes. British Standards Institution, 1998.

**BS EN 737-1:1998.** Medical gas pipeline systems. Terminal units for compressed medical gases and vacuum. British Standards Institution, 1998.

**BS EN 737-2:1998.** Medical gas pipeline systems. Anaesthetic gas scavenging disposal systems. Basic requirements. British Standards Institution, 1998.

**BS EN 737-3:2000.** Medical gas pipeline systems. Pipelines for compressed medical gases and vacuum. British Standards Institution, 2000.

**BS EN 737-4:1998.** Medical gas pipeline systems. Terminal units for anaesthetic gas scavenging systems. British Standards Institution, 1998.

**BS EN 738-2:1999.** Pressure regulators for use with medical gases. Manifold and line pressure regulators. British Standards Institution, 1999.

**BS EN 739:1998.** Low-pressure hose assemblies for use with medical gases. British Standards Institution, 1998.

**BS EN 740:1999, BS 5724-2.204:1999.** Anaesthetic workstations and their modules. Particular requirements. British Standards Institution, 1999.

**BS EN 837-1:1998.** Pressure gauges. Bourdon tube pressure gauges. Dimensions, metrology, requirements and testing. British Standards Institution, 1998.

**BS EN 1044:1999.** Brazing. Filler metals. British Standards Institution, 1999.

**BS EN 1057:1996.** Copper and copper alloys. Seamlesss, round copper tubes for water and gas in sanitary and heating applications. British Standards Institution, 1996.

**BS EN 1254-1:1998.** Copper and copper alloys. Plumbing fittings. Fittings with ends for capillary soldering or capillary brazing to copper tubes. British Standards Institution, 1998.

**BS EN 1412:1996.** Copper and copper alloys. European numbering system. British Standards Institution, 1996.

**BS EN 13348:2001.** Copper and copper alloys. Seamless, round copper tubes for medical gases or vacuum. British Standards Institution, 2001.

**BS EN 60051-1:1999, IEC 60051-1:1997.** Direct acting indicating analogue electrical measuring instruments and their accessories. Definitions and general requirements common to all parts. British Standards Institution, 1999.

**BS EN 60529:1992.** Specification for degrees of protection provided by enclosures (IP code). British Standards Institution, 1992.

**BS EN 60898-1:2003.** Circuit-breakers for overcurrent protection for household and similar installations. Circuit-breakers for ac operation. British Standards Institution, 2003.

**BS EN 60947-2:2003.** Specification for low-voltage switchgear and controlgear. Circuit-breakers. British Standards Institution, 2002.

**BS EN ISO 407:2004.** Small medical gas cylinders. Pin-index yoke-type valve connections. British Standards Institution, 2004.

**BS EN ISO 3549:2002.** Zinc dust pigments for paints. Specifications and test methods. British Standards Institution, 2002.

**BS EN ISO 4126-1:2004.** Safety devices for protection against excessive pressure. Safety valves. British Standards Institution, 2004.

**BS EN ISO 9001:2000.** Quality management systems. Requirements. British Standards Institution, 2000.

**BS EN ISO 11197:2004.** Medical supply units. British Standards Institution, 2004.

**BS EN ISO 13485:2003.** Medical devices. Quality management systems. Requirements for regulatory purposes. British Standards Institution, 2003.

**BS EN ISO 15001:2004.** Anaesthetic and respiratory equipment. Compatibility with oxygen. British Standards Institution, 2004.

**BS ISO 5011:2000.** Inlet air cleaning equipment for internal combustion engines and compressors. Performance testing. British Standards Institution, 2000.

**ISO 32:1977.** Gas cylinders for medical use. Marking for identification of content. International Organization for Standardization, 1977.

**ISO 5145:2004.** Cylinder valve outlets for gases and gas mixtures. Selection and dimensioning. International Organization for Standardization, 2004.

**ISO 7396-1:2002.** Medical gas pipeline systems. Pipelines for compressed medical gases and vacuum. International Organization for Standardization, 2002.

**ISO/DIS 7396-2:2005.** Medical gas pipeline systems. Anaesthetic gas scavenging disposal systems. CEN/TC 215.

## Department of Health publications

**Model Engineering Specification C11 – Medical gases.** HMSO. 1995.

**Model Engineering Specification C51 – Electrical requirements for specified equipment.** HMSO, 1997.

### Firecode

**Health Technical Memorandum 81 – Fire precautions in new hospitals.** HMSO, 1996.

**Health Technical Memorandum 82 – Alarm and detection systems.** HMSO, 1996.

### Health Building Notes (HBNs)

**HBN 26 (Volume 1) – Facilities for surgical procedures.** The Stationery Office, 2004.

### Health Technical Memoranda (HTMs)

**Health Technical Memorandum 2007 – Electrical services supply and distribution.** HMSO, 1993. (issued in four parts)

**Health Technical Memorandum 2011 – Emergency electrical services.** HMSO, 1993. (issued in 4 parts)

**Health Technical Memorandum 2014 – Abatement of electrical interference.** HMSO, 1993. (issued in 4 parts)

**Health Technical Memorandum 2015 – Bedhead services.** HMSO, 1994–95. (issued in 3 parts)

**Health Technical Memorandum 2022 (Supplement 1) – Dental compressed air and vacuum systems.** The Stationery Office, 2002

**Health Technical Memorandum 2025 – Ventilation in healthcare premises.** HMSO, 1994. (issued in 4 parts)

### Other Department of Health publications

**Advice on the implementation of the Health & Safety Commission's occupational exposure standards for anaesthetic agents.** Department of Health, 1996.

## Miscellaneous publications

**European Pharmacopoeia (Ph. Eur.) 2005 (5th edition).** European Directorate for the Quality of Medicines, 2005.
http://www.pheur.org

Health & Safety Executive (1996). **Anaesthetic agents: controlling exposure under COSHH.** HSE Books, 1996.

Health & Safety Executive (1996). **EH40/2002 – Occupational exposure limits 2002.** HSE Books, 2002.

British Compressed Gases Association (BCGA) (2000). **Code of Practice 30 (CP30): The safe use of liquid nitrogen dewars up to 50 litres.** BCGA, 2000.

British Compressed Gases Association (BCGA) (2002). **Code of Practice 19 (CP19): Bulk liquid oxygen storage at users' premises.** Revision 3. BCGA, 2002.

European Commission. **Directive 2001/83/EC of the European Parliament and of the Council of 6 November 2001 on the Community code relating to medicinal products for human use.** *Official Journal of the European Communities.* 28.11.2001, L 311/67.
http://europa.eu.int/eur-lex/pri/en/oj/dat/2001/l_311/l_31120011128en00670128.pdf

Medical gas pipe systems. Design and installation of QAS 3720.1/206). BSI/Department of Health, 1988.